Embedded Systems Design using the TI MSP430 Series

Embedded Systems Design using the TI MSP430 Series

by Chris Nagy

AMSTERDAM • BOSTON • HEIDELBERG • LONDON
NEW YORK • OXFORD • PARIS • SAN DIEGO
SAN FRANCISCO • SINGAPORE • SYDNEY • TOKYO

Newnes is an imprint of Elsevier Science.

Recognizing the importance of preserving what has been written, Elsevier Science prints its books on acid-free paper whenever possible.

Library of Congress Cataloging-in-Publication Data

Nagy, Chris.
Embedded system design using the TI MSP430 series / Chris Nagy.
p. cm.
Includes index.
ISBN: 0-7506-7623-X
1. Embedded computer systems—Design and construction—Data processing.
2. Texas Instruments MSP430 series microprocessors. I. Title.

TK7895.E42N34 2003
004.2'56—dc21

2003054143

British Library Cataloguing-in-Publication Data
A catalogue record for this book is available from the British Library.

The publisher offers special discounts on bulk orders of this book.
For information, please contact:

Manager of Special Sales
Elsevier Science
200 Wheeler Road
Burlington, MA 01803
Tel: 781-313-4700
Fax: 781-313-4880

For information on all Newnes publications available, contact our World Wide Web home page at: http://www.newnespress.com

10 9 8 7 6 5 4 3 2 1

Printed in the United States of America

Contents

About the Author

Chris Nagy is a senior firmware engineer currently working in the field of utility telemetry systems. He has spent the past several years designing with TI's family of microcontroller products.

What's on the CD-ROM?

Included on the accompanying CD-ROM:

- A fully searchable eBook version of the text in Adobe pdf format
- The code from the design examples in this book
- Various useful documents related to the TI MSP430 family

CHAPTER 1

Introduction

As a developer, I grew up on the PIC. Very seldom will you find an embedded developer who has not used this workhorse at one point in their career. Like others, I knew the parts, the tools, and the company. My university used the 8051 in the class on microcontrollers, but when I got out, I learned the PIC from another developer who had been using it for years. When a new project cropped up, I would dig out the same development tools and methods that I had always relied on.

A few years back, I was forced out of my rut. The new project that I was on had much tighter requirements than I had previously experienced. Downloadable firmware, insanely low current budgets and the omnipresent "Don't spend too much" requirements added up to one inescapable fact: my old standby wasn't going to cut it this time. I needed a better microcontroller.

One of my fellow developers introduced me to the Texas Instruments MSP430. If the marketing literature and datasheet were correct, it would do everything I needed. However, as this book discusses later, I seldom trust datasheets, let alone those brightly colored marketing brochures that I seem to receive from every semiconductor manufacturer under the sun. So, I ordered a development kit and played with it. It proved (and has continued to prove) to be a very powerful and useful device.

However, one thing the PIC (and the 8051, and the Motorola HC series, and the Atmel AVR…) has over the MSP430 is available literature. The PIC has at least a dozen books written about development. When I tried to find a book for the '430 family, I came up empty. This is intended to be the book I couldn't find. I hope you find it useful.

About this book

This book is intended for the embedded engineer who is new to the field, and as an introduction and reference for those experienced with microcontroller development, but are new to the MSP430 family of devices. I have assumed that the reader has some experience with microcontroller-based design, either professionally or academically. As an example, the book describes interrupt functionality in detail, but assumes that you, the reader, already know what an interrupt is and how to use it. It is also important to note that, while much of the information in this book is identical to that which is available from the TI documentation, this book is intended to supplement, not replace that valuable source of information. The Users Guides and Application Notes together offer a depth and breadth of technical information that would be difficult to replicate in a single source. The intent of this book is to highlight some of the most commonly used information, along with some (hopefully) helpful suggestions and rules of thumb.

Examples provided in this book are of mixed type, in that some are developed in assembly language, and some in C. The overwhelming majority of small to medium sized embedded projects are developed in C, with good reason. C offers the best balance between easy to develop, structured, portable, readable code and direct control of the hardware (see *Development Language Selection*, pg. 154). I have assumed that the reader is experienced with C language programming

The book is divided into three sections. The first section consists of chapters 2 through 7, and describes the devices themselves. Much of this material is also available from the TI user's guides, which are available for download. I have attempted to develop some objective suggestions and rules of thumb, based on my experiences and those of other developers. My intent is to supplement the user's guide and datasheet, not to replace them.

The second section, which consists of chapters 8 and 9, discusses MSP430 instruction set and addressing. Although use of a higher level language such as C will obscure this level of development, it is still important material. In any project of significant size, the developer will, at some point, need to optimize the code. That developer can either trust the compiler to optimize everything, or become very comfortable with the material in this section. I recommend the latter.

The last four chapters form the third section, which combines information from the earlier chapters and outlines some system development guidelines. Much of what is included in this section (particularly Chapter 11) is not specific to the MSP430, but is useful nonetheless. The section concludes with a simple application. Appendices to the book include a list of some useful resources, a glossary of common acronyms, and a description of the use of the flash emulation tool, a low-end prototyping board from TI.

The structure of this book is, to some degree, circular. Code examples in the first section assume knowledge of the information in the second section, and vice versa. However, the material is straightforward enough that, with some basic knowledge of microcontrollers (along with some use of the index), the reader should not experience any major difficulties.

The MSP430 Family

The MSP430 family is a broad family of low power, feature rich 16-bit microcontrollers from Texas Instruments. They share a common, RISC-type, von Neumann CPU core (See *Architecture Types*, pg. 7). The '430 is competitive in price with the 8-bit controller market, and supports both 8 and 16-bit instructions, allowing migration from most similarly sized platforms.

The family of devices ranges from the very small (1k ROM, 128 bytes for RAM, sub-dollar) up to larger (60k ROM, 2k RAM, with prices in the $10 range) devices. Currently, there are at least 40 flavors available, with more being added regularly. The devices are split into three families: the

MSP430x3xx, which is a basic unit, the MSP430x1xx, which is a more feature-rich family, and the MSP430x4xx, which is similar to the '1xx, with a built in LCD driver. You will find these referred to as '1xx, '3xx, and '4xx devices throughout this book.

Part Numbering Convention

Part numbers for MSP430 devices are determined based on their capabilities. All device part numbers follow the following template:

MSP430$M_tF_aF_bM_c$

where:

M_t: Memory Type

- C: ROM
- F: Flash
- P: OTP
- E: EPROM (for developmental use. There are few of these.)

F_a, F_b: Family and Features

- 10, 11: Basic
- 12, 13: Hardware UART
- 14: Hardware UART, Hardware Multiplier
- 31, 32: LCD Controller
- 33: LCD Controller, Hardware UART, Hardware Multiplier
- 41: LCD Controller
- 43: LCD Controller, Hardware UART
- 44: LCD Controller, Hardware UART, Hardware Multiplier

M_c: Memory Capacity

- 0: 1kb ROM, 128b RAM
- 1: 2kb ROM, 128b RAM
- 2: 4kb ROM, 256b RAM
- 3: 8kb ROM, 256b RAM

4: 12kb ROM, 512b RAM
5: 16kb ROM, 512b RAM
6: 24kb ROM, 1kb RAM
7: 32kb ROM, 1kb RAM
8: 48kb ROM, 2kb RAM
9: 60kb ROM, 2kb RAM

Example: The MSP430F435 is a Flash memory device with an LCD controller, a hardware UART, 16 kb of code memory, and 512 bytes of RAM.

The part numbering scheme described above is a bit fragmented. There are common features not consistently represented (type of ADC, number of timers, etc), and there are some other inconsistencies (for example, the 33 family has the multiplier, but the 13 and 43s do not). I would recommend against selecting parts based on their numbering scheme. Rather, once you have a vague idea of your requirements, go to the TI website (*www.TI.com*), and use their parametric sort feature.

Writing Code

Now that you have selected a device, it is time to begin writing some software for your application. I have several rules for developing code, which I have listed here. These are undoubtedly things you already know, as they are pretty basic and obvious, but it is surprising how often, in the middle of a project, with deadlines bearing down, these are forgotten:

1) *Be consistent.* Develop a standard set of rules for naming constants, variables, and function calls, and stick to it. Later on, when you are reading or editing some remote piece of code, it is handy to immediately recognize that DataOffset is a variable, while DATAOFFSET is a program constant.

2) *Comment your code.* If you can write code that will be completely problem free, and are certain, without a doubt, that the code will never need to be read, changed, or reused for another application, you can ignore this suggestion. However, for us mortal programmers here in the real world, this rule is important. (This is the one that I

tend to let slip. I expect several of my coworkers to mention this hypocrisy to me when they read this.)

3) *Develop on paper.* The best code is scribbled out on paper first, and typed in later. This gives you several opportunities to think about the algorithms and implementation. It is also much easier to perform any necessary analysis, like timing or memory use, on paper, than it is on the fly while typing. I do all of my work in a lab notebook first, allowing me to go back and refer to it later.

4) *Use other eyes.* Code reviews with other developers are the norm in industry, but I believe that they are among the most misused of software development processes. Often, we have a box to check off in our formal processes, so we gather 3 or 4 fellow software guys, e-mail them the 10,000 lines of code the night before the review, spend a couple of hours talking about high-level concepts, and call it done. As complete wastes of time go, this is impressive. If, however, you would like the code review to be useful, the process should encourage the others in the review to pick your code apart. Code reviews should always be performed at the function level, and should include hardware and/or system people, depending on what area the function being reviewed affects. Make certain the reviewers have access to the code several days prior to the review, so that they have plenty of opportunity to look it over. The point of the review is to improve the code, not to check off a box on some document.

Every project seems to begin with these rules being followed. How often, though, do we find ourselves, as crunch time nears, hacking away at the PC, just making the thing work? It has been my experience that most of my really stupid coding mistakes have been made when getting away from these rules, particularly the last two. Last-minute hacking and marathon coding and debugging sessions will, in the long run, require more time and produce inferior code compared to well planned and organized development. Maintain the self discipline to follow these guidelines (or your own), and your code quality will improve.

Architecture Types

Microprocessors and microcontrollers are often described in terms of their architecture types. The MSP430 is described as being a RISC processor. It is also described as utilizing a von Neumann architecture. These two descriptions are bandied about often. (In fact, the RISC category is used by so many semiconductor manufacturers in the marketing of their respective microcontrollers that it is approaching buzzword status.) However, many of us breeze over these bits of information when selecting a micro for our application. It can often be helpful to have some understanding of the implications that arise from these definitions.

RISC vs. CISC Architectures

CISC is an acronym for Complex Instruction Set Computing. CISC machines are characterized by variable length instructions, resulting in complicated instruction decoding. Many CISC processors require microcoding to perform the decode tasks. The range of clock cycles required to execute tends to vary broadly from one instruction to the next in CISC processors. In the 8086, for example, shift and rotate instructions require 2 clock cycles, while an integer multiply requires a minimum of 80.

RISC stands for Reduced Instruction Set Computing. As the name suggests, the instruction set in RISC machines has been stripped down to the basics, to facilitate easier instruction decoding, and faster processing. RISC instructions for a given processor are typically fixed in size, similar in format, and all execute in more or less the same number of cycles. (An exception to this is jump and branch instructions in pipelined processors, which typically require an extra cycle or two.) In short, CISC is designed to accomplish as much as possible with each instruction, and RISC is designed to use simple instructions, and jam them through as fast as possible.

The MSP430 is advertised as a RISC processor. This has been the subject of some friendly debate. A computer science purist of my acquaintance has suggested that the '430 is not a true RISC machine, since the instruction length varies for different addressing modes. This condition is a result of instruction set orthogonality (for a discussion of orthogonality, see pg. 113). After discussions with The Purist, I have come to two conclusions about this. First, he is probably right. The MSP430 doesn't meet the strict, narrow definition of RISC processors. The MSP430 resides in the gray area between RISC and CISC, and does a pretty good job of pulling the strengths out of both. Second, it isn't that important. As long as the developer is cognizant of the variation in cycles per instruction (anywhere from 1 to 5 master clock cycles, depending on addressing mode), this device can and should be regarded as a RISC processor.

RISC and CISC are not the only processor architecture philosophies out there. In recent years, approaches such as VLIW (Very Long Instruction Word) and EPIC (Explicitly Parallel Instruction Computing) have begun to take hold in the high-end processor market. However, it will likely be a while before those of us living in the small micro-controller market need to concern ourselves with these.

Harvard vs. von Neumann Architectures

The terms Harvard and von Neumann describe the memory structure of the device. Harvard architectures have separate address spaces for code memory (ROM, Flash, etc.) and for data memory (RAM). In von Neumann devices, code and data are mapped to a single space.

Due to their ability to simultaneously pull instructions from ROM and data from RAM, Harvard architectures are almost always faster than von Neumann architectures. The tradeoff for this speed is flexibility. Harvard architectures require special instructions in order to write to flash blocks, while von Neumann machines are able to rely on their base instruction set to perform these functions. Von Neumann devices are also able to run code out of RAM, which is necessary for implementation of downloadable firmware. (Reprogramming of flash memory is discussed in Chapter 10.)

Most microcontrollers available today are Harvard architectures, as are the majority of small DSPs. The MSP430, which is a true von Neumann device, is among the few exceptions. Code memory, RAM, information memory, special function registers, and interrupt vectors are all mapped into a contiguous 16-bit addressable memory space. While not very fast, the extra flexibility created by the memory addressing makes the MSP430 a powerful and unique device.

CHAPTER 2

Architecture: CPU and Memory

As discussed in chapter 1, the MSP430 utilizes a 16-bit RISC architecture, which is capable of processing instructions on either bytes or words. The CPU is identical for all members of the '430 family. It consists of a 3-stage instruction pipeline, instruction decoding, a 16-bit ALU, four dedicated-use registers, and twelve working (or scratchpad) registers. The CPU is connected to its memory through two 16-bit busses, one for addressing, and the other for data. All memory, including RAM, ROM, information memory, special function registers, and peripheral registers are mapped into a single, contiguous address space.

This architecture is unique for several reasons. First, the designers at Texas Instruments have left an awful lot of space for future development. Almost half the Status Register remains available for future growth, roughly half of the peripheral register space is unused, and only six of the sixteen available special function registers are implemented.

Second, there are plenty of working registers. After years of having one or two working registers, I greatly enjoyed my first experience with the twelve 16-bit CPU scratchpads. The programming style is slightly different, and can be much more efficient, especially in the hands of a programmer who knows how to use this feature to its fullest.

Third, this architecture is deceptively straightforward. It is very flexible, and the addressing modes are more complicated than most other small processors. But, beyond that, this architecture is simple, efficient and clean. There are two busses, a single linear memory space, a rather vanilla processor core, and all peripherals are memory-mapped.

CPU Features

The ALU

The '430 processor includes a pretty typical ALU (arithmetic logic unit). The ALU handles addition, subtraction, comparison and logical (AND, OR, XOR) operations. ALU operations can affect the overflow, zero, negative, and carry flags. The hardware multiplier, which is not available in all devices, is implemented as a peripheral device, and is not part of the ALU (see Chapter 6).

Working Registers

The '430 gives the developer twelve 16-bit working registers, R4 through R15. (R0 through R3 are used for other functions, as described later.) They are used for register mode operations (see Addressing Modes, Chapter 8), which are much more efficient than operations which require memory access. Some guidelines for their use:

- Use these registers as much as possible. Any variable which is accessed often should reside in one of these locations, for the sake of efficiency.
- Generally speaking, you may select any of these registers for any purpose, either data or address. However, some development tools will reserve R4 and R5 for debug information. Different compilers will use these registers in different fashions, as well. Understand your tools.

- Be consistent about use of the working registers. Clearly document their use. I have code, written about 8 months ago, that performs extensive operations on R8, R9, and R15. Unfortunately, I don't know today what the values in R8, R9 and R15 represent. This was code I wrote to quickly validate an algorithm, rather than production code, so I didn't document it sufficiently. Now, it is relative gibberish. Don't let this happen to you. No matter how obvious or trivial register use seems, document it anyway.

Constant Generators

R2 and R3 function as constant generators, so that register mode may be used instead of immediate mode for some common constants. (R2 is a dual-use register. It serves as the Status Register, as well.) Generated constants include some common single-bit values (0001h, 0002h, 0004h, and 0008h), zero (0000h), and an all 1s field (0FFFFh). Generation is based on the W(S) value in the instruction word, and is described by the table below.

W(S)	Value in R2	Value in R3
00	——	0000h
01	(0) (absolute mode)	0001h
10	0004h	0002h
11	0008h	0FFFFh

Program Counter

The Program Counter is located in R0. Since individual memory location addresses are 8-bit, but all instructions are 16 bit, the PC is constrained to even numbers (i.e. the LSB of the PC is always zero). Generally speaking, it is best to avoid direct manipulation of the PC. One exception to this rule of thumb is the implementation of a switch, where the code jumps to a spot, dependent on a given value. (I.e., if value=0, jump to location0, if value=1, jump to location1, etc.) This process is shown in Example 3.1.

Example 3.1 Switch Statement via Manual PC Control

```
    mov   value,R15      ; Put the Switch value into R15
    cmp   R15,#8         ; range checking
    jge   outofrange     ; If R15>7, do not use PC Switch
    cmp   #0,R15         ; more range checking
    jn    outofrange
    rla   R15            ; Multiply R15 by two, since PC is always even
    rla   R15            ; Double R15 again, since symbolic jmp is 2
                           words long
    add   R15,PC         ; PC goes to proper jump
    jmp   value0
    jmp   value1
    jmp   value2
    jmp   value3
    jmp   value4
    jmp   value5
    jmp   value6
    jmp   value7
outofrange
    jmp   RangeError
```

This is a relatively common approach, and most C compilers will implement switch statements with something similar. When implementing this manually (i.e., in assembly language), the programmer needs to keep several things in mind:

- Always do proper range checking. In the example, we checked for conditions outside both ends of the valid range. If this is not performed correctly, the code can jump to an unintended location.
- Pay close attention to the addressing modes of the jump statements. The second doubling of R15, prior to the add statement, is added because the jump statement requires two words when symbolic mode addressing is used.

- Be careful that none of your interrupt handlers have the potential to affect your value register (R15 in the example). If the interrupt handler needs to use one of these registers, the handler needs to store the value to RAM first. The most common procedure is to push the register to the stack at the beginning of the ISR, and to pop the register at the end of the ISR. (See Example 3.2.)

Example 3.2 Push/Pop Combination in ISR

```
Timer_A_Hi_Interrupt
     push   R12             ; We will use R12
     mov    P1IN,R12        ; use R12 as we please
     rla    R12
     rla    R12
     mov    R12,&BAR        ; Done with R12
     pop    R12             ; Restore previous value to R12
     reti                   ; return from interrupt
            :
            :
     ORG    0FFF0h
     DW     Timer_A_Hi_Interrupt
```

Status Register

The Status Register is implemented in R2, and is comprised of various system flags. The flags are all directly accessible by code, and all but three of them are changed automatically by the processor itself. The 7 most significant bits are undefined. The bits of the SR are:

- The Carry Flag (C)
  Location: SR(0) (the LSB)
  Function: Identifies when an operation results in a carry. Can be set or cleared by software, or automatically.
  1=Carry occurred
  0=No carry occurred

- The Zero Flag (Z)
  Location: SR(1)
  Function: Identifies when an operation results in a zero. Can be set or cleared by software, or automatically.
  1=Zero result occurred
  0=Nonzero result occurred

- The Negative Flag (N)
  Location: SR(2)
  Function: Identifies when an operation results in a negative. Can be set or cleared by software, or automatically. This flag reflects the value of the MSB of the operation result (Bit 7 for byte operations, and bit 15 for word operations).
  1=Negative result occurred
  0=Positive result occurred

- The Global Interrupt Enable (GIE)
  Location: SR(3)
  Function: Enables or disables all maskable interrupts. Can be set or cleared by software, or automatically. Interrupts automatically reset this bit, and the reti instruction automatically sets it.
  1=Interrupts Enabled
  0=Interrupts Disabled

- The CPU off bit (CPUOff)
  Location: SR(4)
  Function: Enables or disables the CPU core. Can be cleared by software, and is reset by enabled interrupts. None of the memory, peripherals, or clocks are affected by this bit. This bit is used as a power saving feature.
  1=CPU is on
  0=CPU is off

- The Oscillator off bit (OSCOff)
  Location: SR(5)
  Function: Enables or disables the crystal oscillator circuit (LFXT1). Can be cleared by software, and is reset by enabled external interrupts. OSCOff shuts down everything, including peripherals. RAM and register contents are preserved. This bit is used as a power saving feature.
  1=LFXT1 is on
  0=LFXT1 is off

- The System Clock Generator (SCG1,SCG0)
  Location: SR(7),SR(6)
  Function: These bits, along with OSCOff and CPUOff define the power mode of the device. See chapter 13 for details.

- The Overflow Flag (V)
  Location: SR(8)
  Function: Identifies when an operation results in an overflow. Can be set or cleared by software, or automatically. Overflow occurs when two positive numbers are added together, and the result is negative, or when two negative numbers are added together, and the result is positive. The subtraction definition of overflow can be derived from the additive definition.
  1=Overflow result occurred
  0=No overflow result occurred

Four of these flags (Overflow, Negative, Carry, and Zero) drive program control, via instructions such as cmp (compare) and jz (jump if Zero flag is set). You will see these flags referred to often in this book, as their function represents a fundamental building block. The instruction set is detailed in Chapter 9, and each base instruction description there details the interaction between flags and instructions. As a programmer, you need to understand this interaction.

Stack Pointer

The Stack Pointer is implemented in R1. Like the Program Counter, the LSB is fixed as a zero value, so the value is always even. The stack is implemented in RAM, and it is common practice to start the SP at the top (highest valid value) of RAM. The push command moves the SP down one word in RAM (SP=SP-2), and puts the value to be pushed at the new SP. Pop does the reverse. Call statements and interrupts push the PC, and ret and reti statements pop the value from the TOS (top of stack) back into the PC.

I have one simple rule of thumb for the SP: leave it alone. Set the stack pointer as part of your initialization, and don't fiddle with it manually after that. As long as you are wary of two stack conditions, the stack pointer manages itself. These two conditions are:

- Asymmetric push/pop combinations. Every push should have a pop. If you push a bunch of variables, and fail to pop them back out, it will come back to haunt you. If you pop an empty stack, the SP moves out of RAM, and the program will fail.
- Stack encroachment. Remember, the stack is implemented in RAM. If your program has multiple interrupts, subroutine calls, or manual pushes, the stack will take up more RAM, potentially overwriting values your code needs elsewhere.

Memory Structure

Special Function Registers

Special function registers are, as you might have guessed, memory-mapped registers with special dedicated functions. There are, nominally, sixteen of these registers, at memory locations 0000h through 000Fh. However, only the first six are used. Locations 0000h and 0001h contain interrupt enables, and locations 0002h and 0003h contain interrupt flags. These are described in Chapter 3.

Locations 0004h and 0005h contain module enable flags. Currently, only two bits are implemented in each byte. These bits are used for the USARTs.

Peripheral Registers

All on-chip peripheral registers are mapped into memory, immediately after the special function registers. There are two types of peripheral registers: byte-addressable, which are mapped in the space from 010h to 0FFh, and word-addressable, which are mapped from 0100h to 01FFh. These are detailed in the memory map at the end of this chapter and further explained in Chapter 6.

RAM

RAM always begins at location 0200h, and is contiguous up to its final address. RAM is used for all scratchpad variables, global variables, and the stack. Some rules of thumb for RAM usage:

- The developer needs to be careful that scratchpad allocation and stack usage do not encroach on each other, or on global variables. Accidental sharing of RAM is a very common bug, and can be difficult to chase down. You need to clearly understand how large your stack will become.

- Be consistent about use. Locate the stack at the very end of the RAM space, and place your most commonly used globals at the beginning.
- Never allocate more scratchpad than you need, and always deallocate as quickly as is reasonable. You can never have too much free RAM.

Boot Memory (flash devices only)

Boot memory is implemented in flash devices only, located in memory locations 0C00h through 0FFFh. It is the only hard-coded ROM space in the flash devices. This memory contains the bootstrap loader, which is used for programming of flash blocks, via a USART module. Use of the bootstrap loader is described in Chapter 10, *Flash Memory*.

Information Memory (flash devices only)

Flash devices in the '430 family have the added feature of information memory. This information memory acts as onboard EEPROM, allowing critical variables to be preserved through power down. It is divided into two 128-byte segments. The first of these segments is located at addresses 01000h through 0107Fh, and the second is at 01080h through 010FFh. Use and reprogramming of information memory is detailed in Chapter 10, *Flash Memory*.

Code Memory

Code memory is always contiguous at the end of the address space (i.e. always runs to location 0FFFFh). So, for 8k devices, code runs from 0E000h to 0FFFFh, and for the 60k devices, the code runs from 01100h to 0FFFFh. All code, tables, and hard-coded constants reside in this memory space.

Interrupt Vectors

Interrupt vectors are located at the very end of memory space, in locations 0FFE0h through 0FFFEh. Programming and use of these are described in detail in Chapter 3.

Memory Map

Memory Address	Description
0FFE0h-0FFFFh	Interrupt Vectors
0FFDFh	End of code space-All devices
0F800h	Start of code space-2K devices
0F000h	Start of code space-4k devices
0E000h	Start of code space-8k devices
0D000h	Start of code space-12k devices
0C000h	Start of code space-16k devices
0A000h	Start of code space-24k devices
08000h	Start of code space-32k devices
04000h	Start of code space-48k devices
01100h	Start of code space-60k devices
010FFh	End of Information Memory: Flash devices except 'F110 and 'F1101
0107Fh	End of Information Memory: 'F110 and 'F1101
01000h	Start of Information Memory: Flash devices only
0FFFh	End of Boot Memory: Flash devices only
0C00h	Start of Boot Memory: Flash devices only
09FFh	End of RAM-2k devices
05FFh	End of RAM-1k devices
03FFh	End of RAM-512 byte devices
02FFh	End of RAM-256 byte devices
027Fh	End of RAM-128 byte devices
0200h	Start of RAM-All devices

01B0h-01FFh	Unused (All devices)
01A0h-01Afh	ADC Control ('1xx and '4xx devices) / Unused ('3xx devices)
0180h-019Fh	Timer B ('1xx devices) / Unused ('3xx and '4xx devices)
0160h-017Fh	Timer A (All devices)
0140h-015Fh	ADC Conversion ('1xx and '4xx devices) / Unused ('3xx devices)
0130h-013Fh	Multiplier (All devices)
0120h-012Fh	Watchdog timer, applicable flash control (All devices)
0110h-011Fh	ADC ('3xx devices) / Unused ('1xx and '4xx devices)
0100h-010Fh	Unused (All devices)
00B0h-00FFh	Unused (All devices)
0090h-00Afh	LCD (Byte addressed, '4xx devices) / Unused ('1xx and '3xx devices)
0080h-008Fh	ADC memory control (Byte addressed, '1xx and '4xx devices) / Unused ('3xx devices)
0070h-007Fh	USART (Byte addressed, All devices)
0060h-006Fh	Unused (All devices)
0050h-005Fh	System Clock (Byte addressable, All devices) / Comparator ('1xx and '4xx devices) / Brownout ('4xx devices) / EPROM and crystal buffer ('3xx devices)
0040h-004Fh	Basic Timer and 8-bit Counter (Byte addressable, '3xx and '4xx devices) / Unused ('1xx devices).
0030h-003Fh	I/O ports 5 and 6 control (Byte addressable, '1xx and '4xx devices) / LCD (Byte addressable, '3xx devices)
0020h-002Fh	I/O ports 1 and 2 control (Byte addressable, All devices)

0010h-001Fh	I/O ports 3 and 4 control (Byte addressable, All devices), I/O port 0 (Byte addressable, '3xx devices)
0006h-000Fh	Unused (All devices)
0005h	Module Enables 2 (Byte Addressable, all devices)
0004h	Module Enables 1 (Byte Addressable, all devices)
0003h	Interrupt Flags 2 (Byte Addressable, all devices)
0002h	Interrupt Flags 1 (Byte Addressable, all devices)
0001h	Interrupt Enables 2 (Byte Addressable, all devices)
0000h	Interrupt Enables 1 (Byte Addressable, all devices)

Memory Types

The MSP430 is available with any one of several different memory types. The memory type is identified by the letter immediately following "MSP430" in the part numbers. (Example: All MSP430Fxxx parts are flash decices).

ROM

ROM devices, also known as masked devices, are identified by the letter "C" in the part numbers. They are strict ROM devices, shipped pre-programmed. They have the advantage of being very inexpensive, and may be the best solution for high-volume designs. However, due to high NRE (non-recurring engineering) costs, masked ROM is only cost-efficient when hundreds of thousands (or more) devices are required. They should also only be used for stable designs. If bugs are found too late in the process, the NRE costs have the potential to be repeated.

OTP

OTP is an acronym for "one time programmable", which pretty well describes the functionality of these devices. Identified by the letter "P" in the part number, OTP parts are a good compromise between

ROM and flash parts. OTPs are shipped blank, and can be programmed at any time. They are typically more expensive than ROM. They also require programming, which can be a hindrance in high-volume manufacturing environments. However, OTPs are ideal for low and medium volume applications, and can be a useful intermediate step when you are still uncertain about the stability of the design.

EPROM

TI offers windowed EPROM versions of several devices, intended for use in development. They are identified by the letter "E" in the part number. These devices are electrically programmable, and UV-erasable. EPROM devices are only available for a few devices, and typically cost on the order of $50 each. They are not intended for production use, but make ideal platforms for emulating ROM devices in development.

Flash

Flash devices, identified by the letter "F" in the part number, have become very popular in the past few years. They are more expensive, but code space can be erased and reprogrammed, thousands of times if necessary. This capability allows for features such as downloadable firmware, and lets the developer substitute code space for an external EEPROM. Chapter 10 is dedicated to flash memory reprogramming.

CHAPTER 3

Reset and Interrupts

The '430 offers numerous interrupt sources, both external and internal. Interrupts are prioritized, with the reset interrupt having the highest priority. This chapter covers the reset sources and conditions in detail, and describes the MSP430 interrupt functionality.

Reset Sources

The '430 uses two separate reset signals, one for hardware and one for software. The hardware reset, which is identified in the literature as POR (power on reset), is generated on initial powerup and when the reset line (RST/NMI) is pulled low. The software reset, identified as PUC (power up clear) is generated by the following conditions:

- Watchdog timer expiration (see Chapter 4).
- Security Key violations, either in the Watchdog timer or Flash memory.
- POR (either powerup or low reset line).

PUC can be forced from software by purposely writing security violations in either the Watchdog or Flash, or by neglecting to "pet the dog", thereby allowing Watchdog expiration.

The resets are seemingly harmless, but can be the source of unexpected trouble. Some things to watch out for:

- POR always calls the reset interrupt vector at 0FFFEh. PUC, however, can call the reset interrupt, or the interrupt vector of the subsystem that generated it (Flash, Watchdog, etc), which is typically at 0FFFCh. Even worse, this behavior changes from one device to another. Consult the data sheet of the specific device you are using, and be prepared to code around some difficulties.
- Make sure the power at the RST/NMI pin is well filtered. It has been my experience that these devices have pretty poor noise susceptibility characteristics. Transients on the RST/NMI pin can cause unwanted resets.
- Troubleshooting an unexpected reset can be tricky. If your system begins skipping to the reset vector periodically (or, even worse, aperiodically), try to eliminate the hardware first. Your best friend in this situation is a good digital oscilloscope. If the RST/NMI pin is clean, look at some of the other system signals before going into the software. I once worked on a device which would regularly fail transient testing, regardless of buffering on the reset pin. It turned out that the critical path for the transient was through the crystal oscillator inputs. In my experience, the vast majority of unexpected resets are caused by hardware issues of one nature or another.
- Once you have eliminated hardware, check your timing. Does the software "pet the dog" often enough? If the Watchdog is reset in a main loop, and the system runs too many interrupts, it might or might not make it back from all those ISRs in time to pet the dog again.
- Are you writing to Flash or the Watchdog in code? Did you intend to?

Reset Condition

Upon a reset signal (POR), the Status Register is reset, and the address in location 0FFFEh is loaded into the Status Register. Peripheral registers all enter their powerup state, which are described later in this book, with the peripheral register descriptions themselves.

The PUC is not as simple. The Status Register is still reset, but the Program Counter is loaded with either the reset vector (0FFFEh), or the PUC source interrupt vector, depending on the source and the device. Dig into the datasheet for specific details. Some peripheral registers are reset by PUC, and some are not. These are also described with the peripheral register descriptions.

One of the common problems found in this level of development is that of branched initialization. Put simply, some applications require different setup to be performed the first time the design is powered up than on subsequent powerup cycles. With flash devices, the solution is simple: select a predefined location (usually in information memory), and clear it to 0x00 after first initialization. On any reset, that location can be checked. If it is zero, branch to the subsequent initialization routine, rather than the first initialization routine. As long as you are careful not to overwrite this location later, this works just fine.

In non-flash devices, some applications do the same thing (although not very reliably) with a RAM location. This works as long as there is sufficient capacitance on the supply pin, and outages are short enough. These devices hold RAM with a trickle of current. There are two problems with this. First, the "enough" described above can be remarkably difficult to predict. Second, too much capacitance on the supply line increases the chances of brownout (see Chapter 7). If your application requires reliable branched initialization, it is probably worth using a flash device, even if just for that.

Interrupts

The '430 offers quite a few interrupt sources. All maskable interrupts are turned off by resetting of the GIE (Global Interrupt Enable) flag in the Status Register. Each maskable interrupt also has an individual enable/disable flag, located in peripheral registers or the individual module.

When an interrupt occurs, the program counter of the next instruction and the status register are pushed to the stack. The SR is then cleared, along with the appropriate interrupt flags if the interrupt is single source. One of the important effects of the SR clearing is the disabling of interrupts, via the reset of the GIE flag. Commonly described as non-reentrant (or non-preemptive) interrupts, the effect of this is that interrupts service routines will not be called from other interrupt service routines unless the GIE bit has been toggled manually. Multiple (peripheral) source flags must be reset manually by the programmer. This functionality is charted in Figure 3.1.

Return to program flow is accomplished by the reti instruction. Reti automatically pops the status register and program counter. If your code jumps to a common point, rather than using the reti instruction, you need to account for these extra items on the stack.

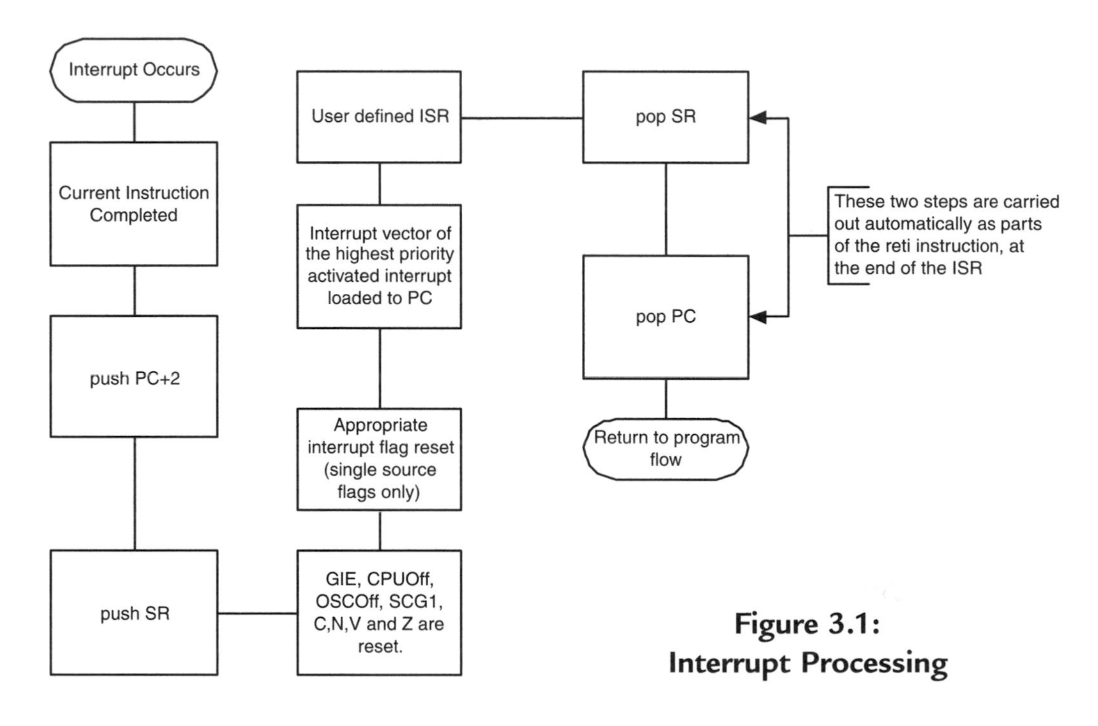

Figure 3.1: Interrupt Processing

Table 3.1: Interrupt Vectors

Vector Address	Priority	'11xx and '12xx	'13x and '14x	'3xx	'4xx
0xFFE0	Lowest	Unused	Unused	Port 0	Basic Timer
0xFFE2	1	Unused	Port 2	Basic Timer	Port 2
0xFFE4	2	Port 1	USART1 Tx	Port 1	USART1 Tx ('44x only)
0xFFE6	3	Port 2	USART1 Rx	Port 2	USART1 Rx ('44x only)
0xFFE8	4	Unused	Port 1	Timer/Port ('32x and '33x)	Port 1
0xFFEA	5	ADC10	Timer_A	ADC ('32x and '33x)/ Timer/Port ('31x)	Timer_A
0xFFEC	6	USART0 Tx ('12xx only)	Timer_A	USART Tx	Timer_A
0xFFEE	7	USART0 Rx ('12xx only)	ADC	USART Rx	ADC ('43x and '44x only)
0xFFF0	8	Timer_A	USART0 Tx	Timer_A	USART0 Tx ('43x and '44x only)
0xFFF2	9	Timer_A	USART Rx	Timer_A	USART0 Rx ('43x and '44x only)
0xFFF4	10	Watchdog Timer	Watchdog Timer	Watchdog Timer	Watchdog Timer
0xFFF6	11	Comparator	Comparator	Unused	Comparator
0xFFF8	12	Unused	Timer_B	Dedicated I/O	Timer_B ('43x and '44x only)
0xFFFA	13	Unused	Timer_B	Dedicated I/O	Timer_B ('43x and '44x only)
0xFFFC	14	Oscillator/Flash/ NMI	Oscillator/Flash/ NMI	Oscillator/Flash/ NMI	Oscillator/Flash/ NMI
0xFFFE	Highest	Hard Reset/Watchdog	Hard Reset/Watchdog	Hard Reset/Watchdog	Hard Reset/Watchdog

Some notes on interrupt sources and vectors:

- The two hardware-dedicated interrupts (Oscillator/Flash and the Hard Reset) vector locations 0xFFFC and 0xFFFE) are non-maskable. All other interrupts are maskable.

- Interrupt Flags may be located in peripheral modules. Consult device data sheets for details.

Many interrupts have multiple flags. Consult device data sheets for details.

Use of Interrupts

Interrupts are the best way to control program flow based on events. Unfortunately, they are also the best way to lose program flow control, based on events. Some guidelines and suggestions for most effective use of interrupts are:

- Always use the reti instruction. If the code manually re-enables interrupts and performs a jump or branch to a common location, the stack will eventually overfill RAM space.
- Make sure all unused interrupts are disabled.
- Generally speaking, your code should not need to manually tinker with the GIE bit via the EINT and DINT instructions. However, do not hesitate to turn interrupts on and off individually. Only keep the locally necessary interrupts enabled at each state of the controller code (e.g. only have the USART interrupts enabled while transmitting or receiving data).
- Plan for the unplanned. Send disabled, and even nonexistent, interrupts to an "Invalid Interrupt" label, with some code to track the interrupt, and a reti instruction. Theoretically, if you followed the previous suggestion, this shouldn't matter that much. Out here in the real world, though, it quite often does. We are all human, and sometimes we humans program incorrect constants into peripheral control registers or interrupt enable flags. If the code tracks and reports which interrupt was triggered, finding the source of the bug can often be done in minutes. If the code doesn't take this into account, the program counter will simply go wandering off the reservation, and your debugging time could stretch into days. This step is almost free, in terms of time and space to program, but it makes your code significantly more robust.

- Another way to handle the unexpected source interrupt is to put only a reti instruction at the Invalid Interrupt label. I recommend against this approach, as it masks the problem, rather than fixing it.
- Be consistent. I begin every new project with the same interrupt vector tables, and modify them as needed. Example 3.1 is my assembly language interrupt vector table for the '11xx family. You might note that even interrupts which are completely undefined by the architecture vector to the Invalid_Interrupt label. There is really no good reason not to take this precautionary step.
- Interrupt handling in C is a bit different. Simply write the ISR (interrupt service routine), and any C compiler worth the CD it is distributed on will handle the vectoring, pushing and popping of CPU registers it uses, and the return instruction. I still use a standard template for interrupts (Example 3.2). A few interesting notes on this example:
    - There is no interrupt routine for the POR (0FFFEh). The compiler should handle this automatically, and vector to your main().
    - All of the interrupt routines consist of an infinite, do-nothing loop. Remember, this is simply a starting point. I will typically write new routines for the interrupts I am using, and leave the endless loop in the others. That way, when I am testing code with an emulator or JTAG debugger, unexpected interrupts become very easy to identify. After the code is performing as expected, I change those eternal loops to something more structured and sensible.
    - The interrupt function structure in this example is for the IAR tools that I use. It can vary from one compiler to the next.

Example 3.1: Assembly Language Interrupt Vector Table ('11xx Family)

```
ORG     0FFE0h
DW      Invalid_Interrupt

ORG     0FFE2h
DW      Invalid_Interrupt

ORG     0FFE4h
DW      Port1_Interrupt

ORG     0FFE6h
DW      Port2_Interrupt

ORG     0FFE8h
DW      Invalid_Interrupt

ORG     0FFEAh
DW      ADC_Interrupt

ORG     0FFECh
DW      Invalid_Interrupt

ORG     0FFEEh
DW      Invalid_Interrupt

ORG     0FFF0h
DW      Timer_A_Hi_Interrupt

ORG     0FFF2h
DW      Timer_A_Lo_Interrupt

ORG     0FFF4h
DW      Watchdog_Timer_Interrupt
```

```
ORG     0FFF6h
DW      Comparator_Interrupt

ORG     0FFF8h
DW      Invalid_Interrupt

ORG     0FFFAh
DW      Invalid_Interrupt

ORG     0FFFCh
DW      Fault_Interrupt

ORG     0FFFEh
DW      Power_Up_Reset
```

Example 3.2: C Language Interrupt Handler Function Template ('14x Family)

```
interrupt [0x02] void dummy_Port2_Interrupt(void)
 {
 while (1)
  {}
 }
interrupt [0x04] void dummy_USART1tx_Interrupt(void)
 {
 while (1)
  {}
 }
interrupt [0x06] void dummy_USART1rx_Interrupt(void)
 {
 while (1)
  {}
 }
```

```
interrupt [0x08] void dummy_Port1_Interrupt(void)
 {
 while (1)
   {}
 }
interrupt [0x0A] void dummy_TimerAH_Interrupt(void)
 {
 while (1)
   {}
 }
interrupt [0x0C] void dummy_TimerAL_Interrupt(void)
 {
 while (1)
   {}
 }
interrupt [0x0E] void dummy_ADC_Interrupt(void)
 {
 while (1)
   {}
 }
interrupt [0x10] void dummy_USART0tx_Interrupt(void)
 {
 while (1)
   {}
 }
interrupt [0x12] void dummy_USART0rx_Interrupt(void)
 {
 while (1)
   {}
 }
interrupt [0x14] void dummy_WDT_Interrupt(void)
 {
 while (1)
```

```
    {}
  }
interrupt [0x16] void dummy_ComparatorA_Interrupt(void)
  {
  while (1)
    {}
  }
interrupt [0x18] void dummy_TimerBH_Interrupt(void)
  {
  while (1)
    {}
  }
interrupt [0x1A] void dummy_TimerBL_Interrupt (void)
{
while (1)
    {}
}
interrupt [0x1C] void dummy_NMI_Oscillator_Interrupt(void)
  {
  while (1)
    {}
  }
```

Guidelines for Interrupt Service Routines

There is really one critical guideline for writing an ISR: KEEP IT SHORT!!!! Overly lengthy ISRs can create a myriad of problems, especially if the routine is intended to be reentrant. If the ISR requires 100 ms to process, and the interrupts are coming every 90 ms, the stack will overflow, and your design will crash, probably quite spectacularly.

My personal rule of thumb is that all ISRs are too long. I am constantly trying to shorten them up. One common practice is to set software flags within the ISR, then return. You can then perform processing in your main function based on those flags. Also remember that any CPU register used in the ISR first needs to be pushed to the stack, which increases your likelihood of overflow.

Common Sources of Error

Stack Overflow You have probably noticed by now that this book keeps issuing warnings about this. This is because this type of error once burned me. The device in question had a particularly critical value stored near the end of useable RAM, and interrupts that tended to come in groups. (Knowing these interrupts would be coming closely together, I set the GIE bit at the beginning of each ISR, in a vain attempt to make my ISRs reentrant. That was another mistake.) There would be a string of interrupts which would produce multiple push operations, which would fill the stack to the point where my critical value had been overwritten, and then pop the stack pointer back to its original location. Debugging required several days. The overflow was a result of ISRs which were too long, a questionable decision about interrupt re-enables, and poor RAM management.

Race Condition Race conditions occur when two interrupts occur very close to each other, and both access the same global variable. Observe Example 3.3, in which timer A is configured to periodically read and accumulate the value from port 3 in the variable foo. An external interrupt on port 1 outputs the value on port 2. This example may produce non-repeatable results when these two interrupts occur very close to each other in time. The problem is further compounded if foo is accessed elsewhere in the code.

Example 3.3: Race Condition Example

```
unsigned int foo=0;
    :
    :
interrupt [0x08] void Port1_Interrupt(void)
 {
 P2OUT=foo;
 }
```

```
interrupt [0x0A] void TimerAH_Interrupt(void)
 {
 foo=foo + P3IN;
 }
```

Boundary Conditions. Boundary conditions are similar to race conditions, and often occur as a result of a race. While not limited to interrupt processing, they most often are a result of a poorly timed interrupt.

An example of this is found in a real time clock application. Time is very often represented as the number of seconds since a predefined standard. Use of a 32-bit value will allow for about 136 years before overflow. In the '430, that value is stored in two 16-bit locations (either in RAM or registers). When a second expires, the low word is incremented. If the low word overflows, the high word is incremented. Between those two increment events, an interrupt can occur. If that interrupt uses the time value, the value that it uses will be off by an order of 2^{16}.

Boundary conditions like this are very rare, but that works against the developer. A bug like this will almost never manifest itself on the bench. More often, it will occur after your design has been sold to the customer. A simple approach can prevent this from occurring. If your ISRs only set flags, and the main program loop then processes based on those flags, boundary conditions are avoidable.

Interrupts vs. Polling

The most common alternative to interrupts for event-based control is polling, which is the process of manually checking values for changes on a repetitive basis, as a part of the main program loop. In college, and early in my career, I learned the two fundamental guidelines for deciding if a given trigger should be polled, or wired as an interrupt. They are:

1. Polling is evil. It is inefficient, and very software intensive. Polling will turn your otherwise short main loop into a lengthy, time consum-

ing one, creating unacceptable system latencies. Interrupts are a better approach.

2. Interrupts are evil. They cause conflicts, require careful stack management, create a host of debugging problems, and always seem to false at the worst possible time. Polling is a better approach.

Obviously, these rules leave a bit to be desired. However, I have heard experienced firmware developers spout each of these opinions as gospel. Both viewpoints make valid arguments.

In my experience the truth lies somewhere in between the two. As in most everything in life, the key to success is balance. Some more useful rules of thumb:

- Use some common sense about the source. Port interrupts lend themselves to polling, while the timers should use interrupts.
- Maintain balance. Most applications I have written use interrupts for the two or three events with the strictest latency requirements, and poll the remaining few event sources.
- Look for conflicts. Many of the error conditions described in this chapter can be avoided with a smart mix of polling and interrupts.
- Understand the timing of the system. Perform worst-case analysis on latency of polled events, and determine if the worst case is adequate. Don't forget to consider the time required for any interrupts which may occur between polled events.

CHAPTER 4

Clocks and Timers

System timing is among the most fundamentally critical areas of embedded design. On most of the systems I have worked on or observed, a significant portion of the debugging time has been dedicated to chasing problems resulting from incorrect or inaccurate clocking. Great care should be expended in timing design, both in hardware and software.

The MSP430 offers multiple clock sources and uses. This chapter is divided into three sections, which discuss clock sources, control, and use. Clock implementations vary significantly among the three device families, so the descriptions in this chapter are categorized with that in mind.

Clock Sources

The '430 devices allow for several different, flexible sources of clock. This area is another prime example of TI's philosophy of giving the developer enough rope to hang themselves. Larger devices offer two independent crystal inputs, along with an internally generated, variable frequency oscillator, that can be divided (or not), and mapped to any of three different internal clocks. The CPU and on-board peripherals can select any of the clocks, and all of the clock signals can be brought out via function pins.

On system reset, the device comes up running off the DCO, at the nominal frequency defined in the device datasheet, with the DCOR bit cleared. This condition also occurs if either the crystal oscillator or high-frequency oscillator is selected as the master clock, or an oscillator fault occurs. This allows the oscillator fault ISR to be processed and, if the system is designed properly, the condition to be diagnosed.

Crystal Oscillator

The crystal oscillator circuit is designed for use with standard 32.768 kHz watch crystals. Depending on the crystal selected, external capacitors may be required. The XIN and XOUT pins have some internal capacitance, which varies by device. Consult the device and crystal datasheets to determine if additional capacitance is required.

- ’1xx Series Crystal Oscillator:

In the ’1xx series, the external crystal oscillator produces an internal signal identified by LFXT1CLK. It can be configured in either low-speed mode (32.768 kHz), or with a high-speed crystal, which can typically be up to 8 MHz in frequency. Crystal frequency range is selected with the XTS bit. The oscillator is enabled by a collection of processor bits. The logic for the enable is shown in Figure 4.1.

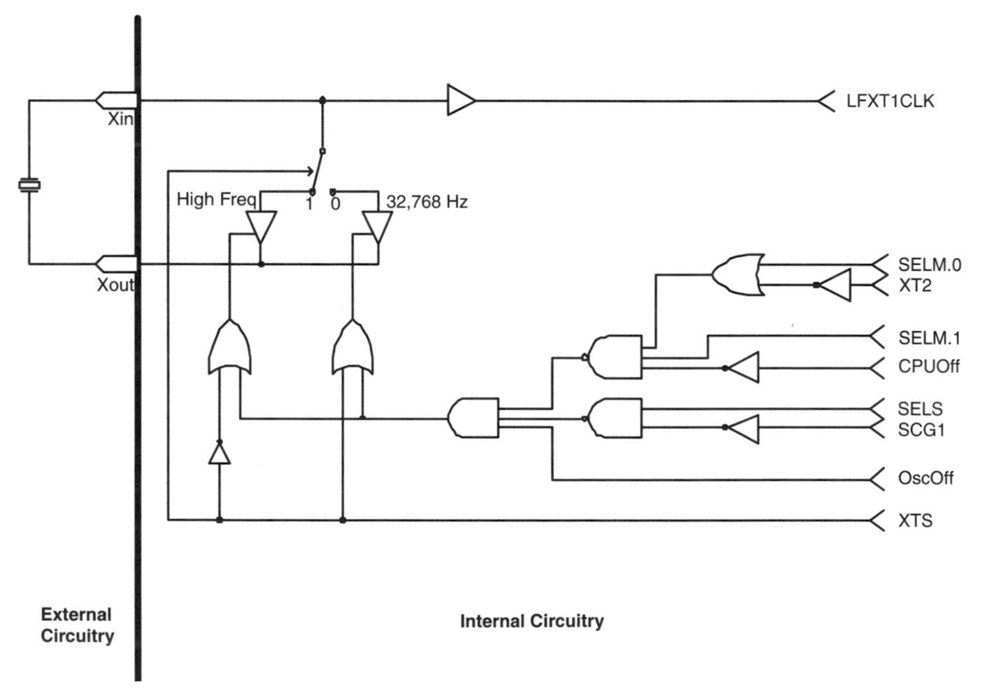

Figure 4.1: MSP430x1xx Crystal Oscillator

An oscillator fault occurs when no cycles occur for a nominal time period (typically about 50 microseconds – as always, consult your datasheet). When this occurs, the NMI/Oscillator Fault interrupt is called. The MCLK is switched to the nominal DCO value, if it was running from LFXT1CLK. It is important to remember that the fault detection circuitry only operates when the crystal oscillator is in high-frequency mode (i.e. XTS=1). Faults occurring when operating in the 32.768 kHz range will go undetected.

- *'3xx Series Crystal Oscillator*

The '3xx series has a very simple crystal oscillator circuit (see Figure 4.2). It produces a single internal signal, ACLK. There is no fault detection, and a single enable bit in the status register, OscOff. The '3xx series crystal oscillator has no high speed mode, only supports 32.768 kHz crystals.

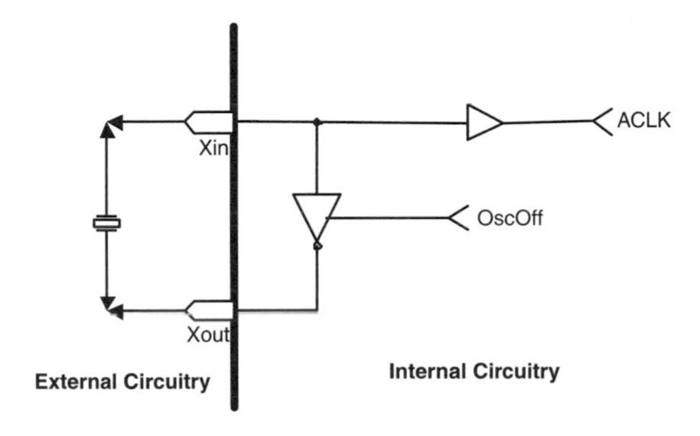

Figure 4.2: MSP430x3xx Crystal Oscillator

- *'4xx Series Crystal Oscillator*

The '4xx series crystal oscillator supports both high-speed (XTS_FLL=1) and low-speed (XTS_FLL=0) crystals.

Figure 4.3: MSP430x4xx Crystal Oscillator

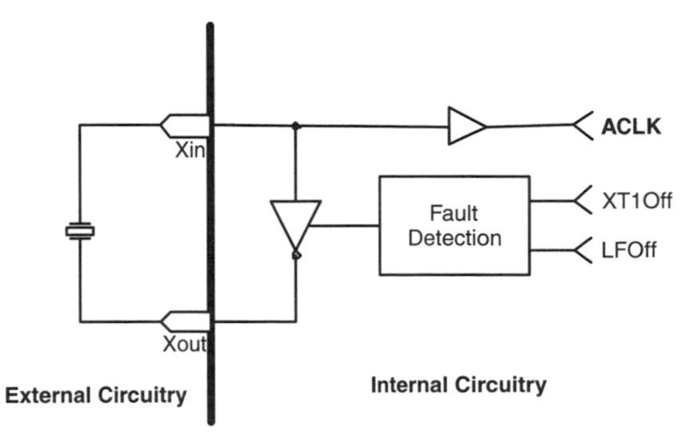

Digitally Controlled Oscillator (DCO)

The DCO is a built-in RC-type oscillator, with a wide, software-controllable frequency range. Clock precision is achieved through use of the Frequency Locked Loop, which is detailed in this section.

- *Frequency Locked Loop (FLL) Operation*

'3xx and '4xx devices offer Frequency Locked Loop, which modulates the DCO frequency, allowing for greater precision and control. It operates by mixing the programmed DCO frequency with the next highest DCO frequency. Each 32 clock cycles are divided into (32-n) cycles at f_{DCO} and n cycles at f_{DCO+1}, where n is a 5-bit value stored in control registers (we will describe the control registers shortly, as they vary by family). There are 29 unique values for n, as the values 28, 29, 30, and 31 produce identical modulation.

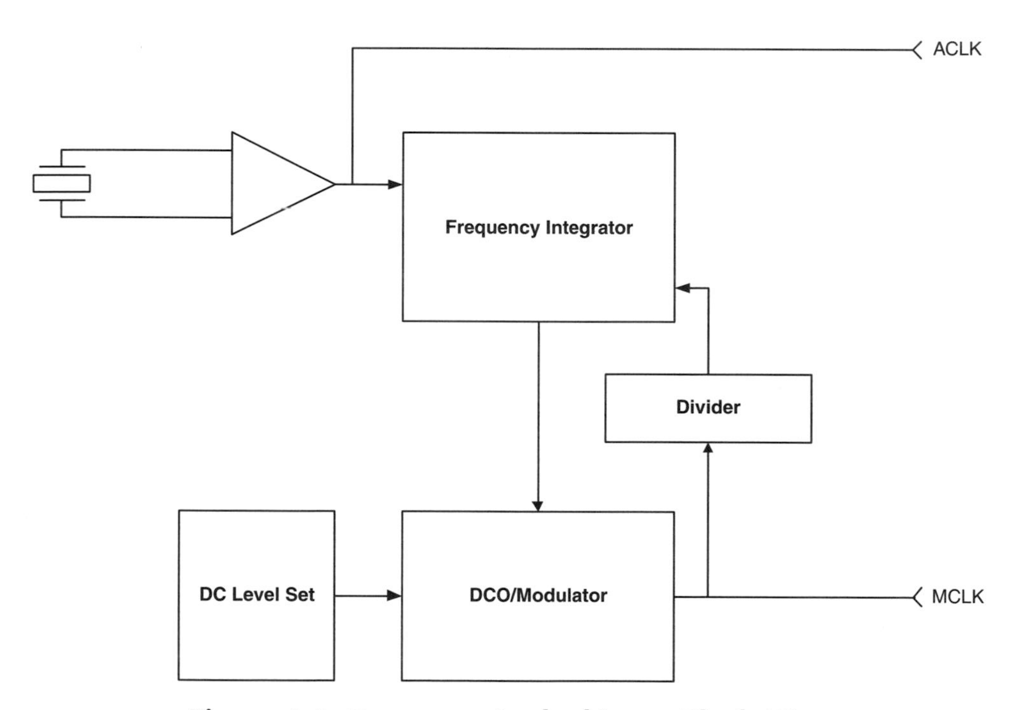

Figure 4.4: Frequency Locked Loop Block Diagram

While this approach is accurate for longer time periods (>> 32 clock cycles), it is inaccurate for shorter timing measurements. The f_{DCO} and f_{DCO+1} frequencies are distributed roughly equally through the cycle of 32 clocks, which helps this inaccuracy slightly, but this is something the developer needs to be wary of. This inaccuracy is more pronounced when there is a larger discrepancy between f_{DCO} and f_{DCO+1} frequencies (e.g. when there are 31 steps at one frequency, and a single step at the other) than when they are evenly balanced (e.g. 16 steps at each frequency).

The solution I have used to handle this inaccuracy is simple: use a crystal to meet short period precision timing requirements. Both the DCO and crystals suffer from error due to aging and temperature. I have found that crystal inaccuracies are usually more predictable than those introduced by the DCO, since they do not add the unpredictability of guessing where in the f_{DCO}/f_{DCO+1} cycle the timing process began.

■ *'1xx Series DCO*

The '1xx series features a DCO which generates an internal signal identified as DCOCLK, that can be programmed either internally or externally. External programming is selected by setting the DCOR bit, and controlled via a resistor connected to the R_{OSC} and V_{CC} pins. When the DCOR bit is cleared, the DCO frequency is controlled internally.

The '1xx family does not have the full FLL functionality. It does, however, offer the frequency modulation method described above.

The DCO control bits are:

- RSEL2, RSEL1, and RSEL0. These bits select the frequency range of the DCO.
- DCO2, DCO1, and DCO0. These bits set the fundamental frequency of the DCO, within the range defined by the RSEL bits.
- MOD4 through MOD0. These are the modulation bits, whose function is described above (FLL Operation).

- Control Bits. See Figure 4.5 for the logical function of various CPU flags in enabling and controlling the DCO.

Specific frequency ranges and values vary by device, and are described in the datasheet for the part you are using.

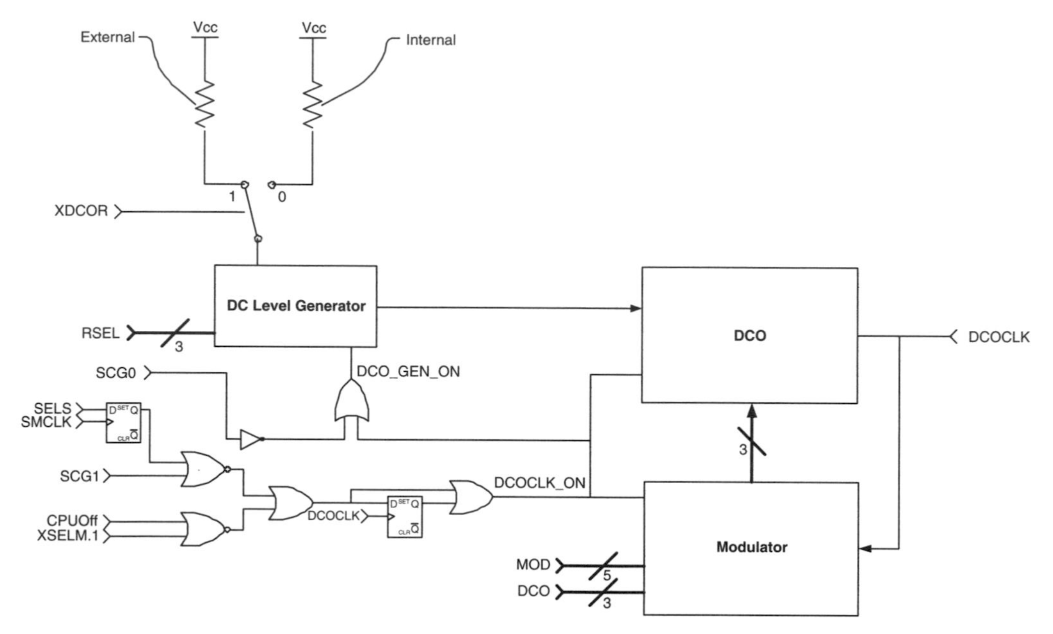

Figure 4.5: '1xx Series Digitally Controlled Oscillator

- *'3xx Series DCO*

The '3xx series DCO is an internally controlled oscillator, which generates a signal identified as MCLK. MCLK is a function of ACLK, which is generated by the crystal oscillator. The relationship is MCLK = (N+1)ACLK, where N is the multiplier defined by the 7 LSBs of the register SCFQCTL. On reset, N is initialized to a value of 31.

The '3xx series uses the same N,N+1 frequency modulation scheme as '1xx devices, with the modulation bits located in the SCFI1 and SCFI0 registers. (The three MSBs are located in SCFI1.2, SCFI1.1, and SCFI1.0, and the two LSBs are SCFI0.1 and SCFI0.0.) In the '3xx series, however, these bits are set automatically, by a 10-bit frequency integrator. The integrator compares MCLK with ACLK(N+1), and sets or clears these bits accordingly. The developer simply needs to control the value N in order to change clock frequency.

- *'4xx Series DCO*

The '4xx series DCO is nearly identical in function and structure to the DCO in the '3xx series. The DCO generates a signal identified as f_{DCOCLK}, which is set equal to ACLK x D x (N+1). Unlike the '3xx, there is a second multiplier, D, which is located in the two MSBs of SCFI0. D is further controlled by the DCO+ bit. Clearing the DCO+ bit removes D from the calculation of f_{DCOCLK}. On reset, D=2, but DCO+ is cleared, giving an effective multiplier of 32.

High-frequency Oscillator

Some of the '1xx series devices allow for operation from two completely independent crystals. The low-frequency crystal input, described above, is designed around 32.768 kHz crystals. The second input, identified as XT2, is designed for use with higher-frequency crystals. The XT2 oscillator behaves identically to the LFXT oscillator in high-frequency mode. It is enabled and disabled using the XT2Off bit, which is generated as shown in Figure 4.6.

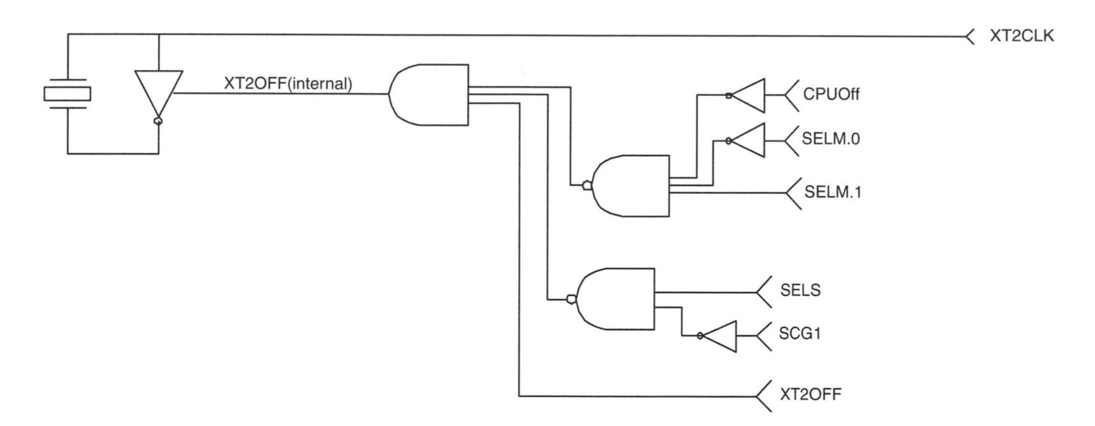

Figure 4.6: High-frequency Oscillator

Clock Controls

'1xx Series Clock Controls

'1xx series devices are controlled by the Basic Clock Module (BCM). The DCO, low-frequency oscillator, and high-frequency oscillator (if available) all act as inputs to the BCM. The BCM controls these sources, divides them down to a lower frequency if desired, and routes them to three available system clock sources:

- MCLK, the Main System Clock. The MCLK can be sourced by any of the three clock sources.
- SMCLK, the Sub-System Clock, or Sub-Main Clock. SMCLK is sourced by either the DCO or the XT2 inputs.
- ACLK, the Auxiliary clock. ACLK is always sourced by the LFXT1CLK source.

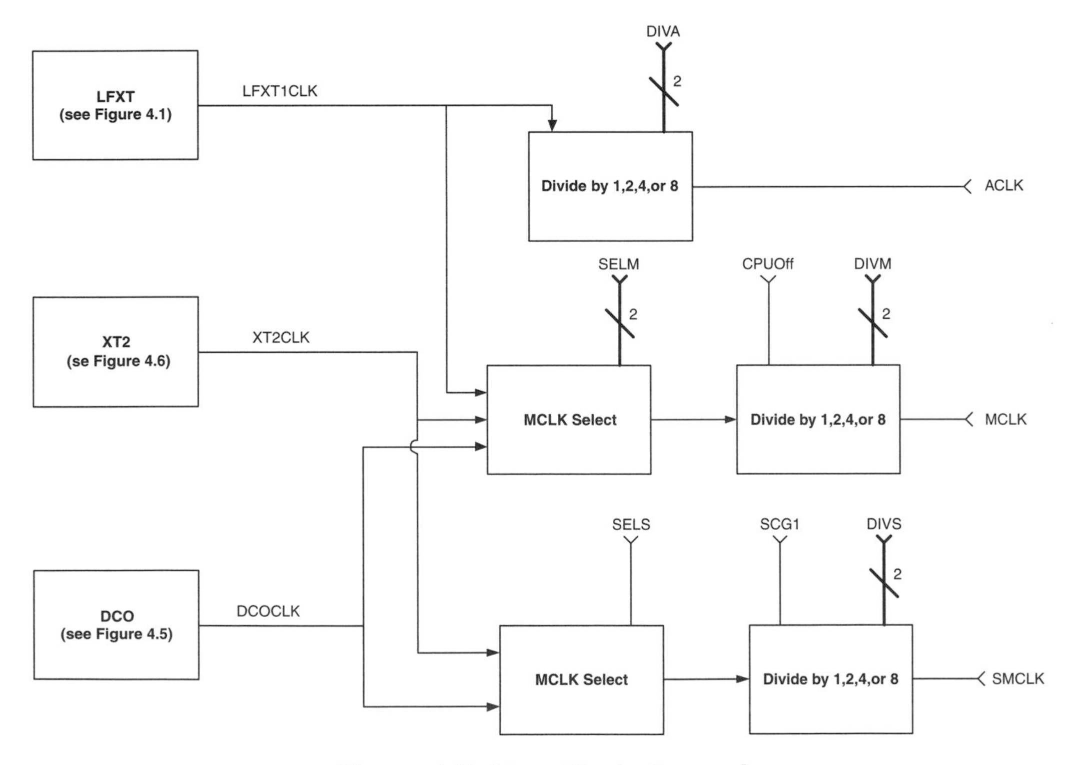

Figure 4.7: '1xx Clock Control

The BCM is configured using three byte-addressable registers, all of which are fully software controllable. Their pertinent details are:

- DCOCTL, the DCO control register.
  Address: 0x0056h
  Structure:

Bit	DCO.2	DCO.1	DCO.0	MOD.4	MOD.3	MOD.2	MOD.1	MOD.0
Reset Value	0	1	1	0	0	0	0	0

DCO (3 MSBs) : Defines the base DCO frequency.
MOD (5 LSBs) : Defines the number of ticks in the N,N+1 modulation.

- BCSCTL1, Oscillator control register #1.
  Address: 0x0057h
  Structure:

Bit	XT2Off	XTS	DIVA.1	DIVA.0	Res*	RSEL.2	RSEL.1	RSEL.0
Reset Value	1	0	0	0	0	1	0	0

*Reserved bit. Do not set this bit.

XT2Off (MSB) : Enables/disables the XT2 Oscillator
XT2Off=0: XT2 is on
XT2Off=1: XT2 is off

XTS : Selects the operating mode for LFXT1 oscillator
XTS=0: Low-Frequency Mode
XTS=1: High-Frequency Mode

DIVA : Selects the divider state for ACLK. ACLK is divided by:
DIVA=0: 1
DIVA=1: 2
DIVA=2: 4
DIVA=3: 8

RSEL(3 LSBs) : Selects one of the eight discrete steps for the DCO.

- BCSCTL2, Oscillator control register #2.
  Address: 0x0058h
  Structure:

Bit	SELM.1	SELM.2	DIVM.1	DIVM.0	SELS	DIVS.1	DIVS.0	DCOR
Reset Value	0	0	0	0	0	0	0	0

SELM (3 MSBs) : Selects the source for MCLK

SELM=0 or 1: MCLK is sourced by the DCOCLK

SELM=2: MCLK is sourced by XT2 (if available)

MCLK is sourced by LFXT1CLK (if no XT2)

SELM=3: MCLK is sourced by

LFXT1CLK

DIVM : Selects the divider state for MCLK. MCLK is divided by:

DIVM=0: 1

DIVM=1: 2

DIVM=2: 4

DIVM=3: 8

SELS : Selects the source for SMCLK

SELS=0: SMCLK is sourced by the DCOCLK

SELS=1: SMCLK is sourced by XT2 (if available)

SMCLK is sourced by LFXT1CLK (if no XT2)

DIVS : Selects the divider state for SMCLK. SMCLK is divided by:

DIVS=0: 1

DIVS=1: 2

DIVS=2: 4

DIVS=3: 8

DCOR (LSB) : Selects the source for the DCO

DCOR=0: Internal DCO resistor is activated

DCOR=1: External Resistor Required

'3xx Series Clock Controls

The '3xx series controls clock signals with two sets of registers. The first set of registers, consisting of BTCTL, BTCNT1 and BTCNT2, control the Basic Timer, which produces low-frequency signals for use by peripheral modules. The second set, which includes SCFQCTL, SCFI0, SCFI1, and CBCTL, are general-purpose controls, which configure system clocks. The details of these registers:

- BTCTL, Basic Timer Control Register.
  Address: 0x0040h
  Structure:

Bit	SSEL	HOLD	DIV	FRFQ.1	FRFQ.0	IP.2	IP.1	IP.0
Reset Value	0	0	0	0	0	0	0	0

SSEL(MSB) and DIV : Selects the source for BTCNT2

SSEL,DIV=00: ACLK is selected

SSEL,DIV=01: ACLK/256 is selected

SSEL,DIV=10: MCLK is selected

SSEL,DIV=11: ACLK/256 is selected

HOLD : Stops the counter operation

BTCNT1 is held if HOLD and DIV are set.

BTCNT2 is held if HOLD is set

FRFQ : Selects the LCD frequency

The LCD frequency is ACLK divided by 32*(1+FRFQ)

IP : Selects the divider value for the interrupt interval timer. This sets the frequency of periodically generated interrupts by the Basic Timer:

The pre-divided ACLK/MCLK (depending on SSEL, DIV) are divided by 2^(IP+1). There is a single exception: for the value IP=001, the clock is divided by 2 (instead of 4, as the above equation would indicate).

- BTCNT1, Basic Timer Counter 1.
  Address: 0x0046h
  Structure:

Bit	CNT1.7	CNT1.6	CNT1.5	CNT1.4	CNT1.3	CNT1.2	CNT1.1	CNT1.0
Reset Value	0	0	0	0	0	0	0	0

BTCNT1 divides ACLK by the value in CNT1, for use in peripheral modules.

- BTCNT2, Basic Timer Counter 2.
  Address: 0x0047h
  Structure:

Bit	CNT2.7	CNT2.6	CNT2.5	CNT2.4	CNT2.3	CNT2.2	CNT2.1	CNT2.0
Reset Value	0	0	0	0	0	0	0	0

BTCNT2 divides the input clock, which is selected by SSEL and DIV (in BTCTL), by the value in CNT2, for use in peripheral modules.

- SCFQCTL, System Clock Control.
  Address: 0x0052h
  Structure:

Bit	M	SCF.6	SCF.5	SCF.4	SCF.3	SCF.2	SCF.1	SCF.0
Reset Value	0	0	0	1	1	1	1	1

M: The modulation bit. If set, adjacent DCO taps are not mixed (i.e., the system frequency is set at a single DCO tap level).

SCF: System clock multiplier. The system clock is equal to the crystal frequency multiplied by (SCF+1).

- SCFI0, System Clock Integrator Control.
  Address: 0x0050h
  Structure:

Bit	<res>	<res>	<res>	FN.4	FN.3	FN.2	INT.1	INT.0
Reset Value	0	0	0	0	0	0	0	0

<res>: Reserved (unused) bits. Read-only

FN.4-FN.2: DCO Frequency range select.

FN=000 MCLK=$f_{nominal}$
FN=001 MCLK=2x$f_{nominal}$
FN=01x MCLK=3x$f_{nominal}$
FN=1xx MCLK=4x$f_{nominal}$

INT.1-INT.0: This 10-bit value, representing the integrator value, is contained in this register, and SCFI1. INT.1 and INT.0 are the two LSBs.

- SCFI1, System Clock Integrator Control.
  Address: 0x0051h
  Structure:

Bit	INT.9	INT.8	INT.7	INT.6	INT.5	INT.4	INT.3	INT.2
Reset Value	0	0	0	0	0	0	0	0

INT.9-INT.2: This 10-bit value, representing the integrator value, is contained in this register and SCFI0. INT.9 through INT.2 *are the eight MSBs*.

- CBCTL, Crystal Buffer Control.
  Address: 0x0053h
  Structure:

Bit	<res>	<res>	<res>	<res>	<res>	CBSEL.1	CBSEL.0	CBE
Reset Value	0	0	0	0	0	0	0	0

<res>: Reserved (unused) bits. Read-only.

CBSEL: Selects the XBUF frequency
- CBSEL=00: ACLK
- CBSEL=01: ACLK/2
- CBSEL=10: ACLK/4
- CBSEL=11: MCLK

CBE: Crystal Buffer Enable
- CBE=0: Crystal Buffer Disabled
- CBE=1: Crystal Buffer Enabled

'4xx Series Clock Controls

The '4xx series clock controls are very similar to those of the '3xx series. The Basic Timer registers, BTCTL, BTCNT1 and BTCNT2, are identical to those of the '3xx, as are SCFQCTL, SCFI0, and SCFI1. CBCTL, however, is not implemented in these devices. Rather, it is replaced by two FLL and control registers, FLL+CTL0 and FLL+CTL1. Their details:

- FLL+CTL0, FLL/Control 0.
  Address: 0x0053h
  Structure:

Bit	DCO+	XTS_FLL	OscCap.1	OscCap.2	XT2OF	XT1OF	LFOF	DCOF
Reset Value	0	0	0	0	0	0	1	1

DCO+: If set, the DCO output is divided, based on FLL_DIV, before sourcing MCLK or SMCLK.

CBE: Set this bit when using a high-frequency crystal for LFXT1.
Clear this bit when using a 32.768 kHz crystal.

OscCap: Internal load capacitance of crystal oscillator.
OscCap=00: Negligible Internal Load Capacitance
OscCap=01: 10 pF (on each crystal pin).
OscCap=10: 14 pF (on each crystal pin).
OscCap=11: 18 pF (on each crystal pin).

XT2OF, XT1OF, LFOF, DCOF:
These read-only flags are used in determination of oscillator fault condition. Do NOT attempt to write these pins.

- FLL+CTL1, FLL/Control 1.
  Address: 0x0054h
  Structure:

Bit	<res>	SMCLK Off	XT2Off	SELM	SELM	SELS	FLL_DIV .1	FLL_DIV .0
Reset Value	0	0	0	0	0	0	0	0

<res>: Reserved (unused) bits. Read-only.
SMCLK_Off: When set, the SMCLK is disabled.
XT2Off: When cleared, the XT2 is disabled.

SELM: Selects the source for MCLK.
SELM=00: DCOCLK sources MCLK
SELM=01: DCOCLK sources MCLK
SELM=10: XT2CLK sources MCLK
SELM=11: LFXT1CLK/ACLK sources MCLK

SELS: Selects the source for SMCLK.
SELS=0: DCOCLK sources SMCLK
SELS=1: XT2CLK sources SMCLK

FLL_DIV: Select the divider value for ACLK.
FLL_DIV=00: ACLK is passed undivided
FLL_DIV=01: ACLK is divided by 2
FLL_DIV=10: ACLK is divided by 4
FLL_DIV=11: ACLK is divided by 8

Clock Uses

CPU Clock

The most basic use of clock is to drive the CPU. Configuration of the CPU clock is described in the sections above, and is typically among the first of the housekeeping items performed on reset. The MSP430 allows for reconfiguration of CPU clock at any time, so the developer may accelerate instruction speed for time-critical operations and slow it down to preserve power. Additionally, the CPU clock can be completely disabled, with only peripheral devices operating. This is detailed in Chapter 12, *Low Power Design Guidelines*.

The relationship between clock speed and instruction speed is a tricky one with this CPU. Instructions can require from one to six clock cycles to process, depending on the instruction itself and the addressing mode. In time critical situations, it is often necessary to manually count cycles. This practice is straightforward enough when the source code is developed in assembly. When source is developed in a compiled language (such as C or FORTH), you will need to generate and examine the disassembled code.

For non-time-critical situations, it is still necessary to know your instruction speed, at least approximately. The general rule of thumb upon which I rely is that for most code, the instruction speed is about one-third of the CPU clock speed. I have found this generalization to be pretty good, typically within about 20%. Smaller systems sometimes provide an exception to this rule, however. When there are very few system variables, it is common to use register mode instructions almost exclusively, and the instruction speed becomes considerably faster than the 1/3rd approximation.

Watchdog Timer

The Watchdog Timer is designed primarily as an error recovery mechanism. It is an independent counter, which, upon overflow, issues an interrupt request. The idea is that, if the software has a problem which causes it to hang, seize, or "wander off the reservation," the watchdog timer will expire and reset the controller. Normally operating code will include periodic reset of the watchdog timer, to prevent this reset when no error condition occurs. This reset is commonly referred to as "petting the dog," or "kicking the dog," depending on the proclivities of the individual describing the process.

The '430 watchdog timer counter is identified here and in the TI literature as WDTCNT. This counter is not accessible by the user. All control of the WDT is performed through the Watchdog Timer Control Register, WDTCTL. WDTCTL is a 16-bit register, functionally split in half. The eight LSBs are described below. The eight MSBs perform a password function. When read, these will always return a value of 0x69h. When writing to WDTCTL, the software must place the value 0x5A into the top half of WDTCTL. Any other value written to that sub-register will cause a system reset. This is actually a very useful feature, as it provides the developer a method to force a PUC from software. I have used this feature in fault-recovery routines more than once.

The watchdog timer operates in one of two modes. The first, interval timer mode, is selected when TMSEL=1. In this mode, the watchdog timer generates a "standard" interrupt (i.e., no PUC) upon WDTCNT overflow. The interrupt can be forced by writing a 1 to CNTCL in this mode. I tend to use this mode in most designs, by writing the ISR so that some important values, such as the processor registers and the top 4 or 5 values in the stack, are written to a pre-defined block in information memory, so they can be retrieved later. I will then change the watchdog mode, and force the hard reset. Doing this provides some extra insight into the state of the processor when the timer expired, since the stack contains the value that was in the PC when the ISR was called. (While the PUC itself does not erase all of these useful values, your startup routine often will.) This little bit of extra code has paid huge dividends when debugging on more than one occasion.

The second mode (TMSEL=0) is the true watchdog timer mode. When the WDTCNT overflows, a PUC is issued. In both modes, the code should periodically pet the dog with the command MOV #05A0Ah,&WDTCTL (in assembly language), or WDTCTL=0x05A5; (in C).

- WDTCTL, Watchdog Timer Control.
  Address: 0x0120h
  Bit 3 is write-only. All other bits are readable and writable.
  Structure:

Bit	PWD.7	PWD.6	PWD.5	PWD.4	PWD.3	PWD.2	PWD.1	PWD.0
Reset Value	0	1	1	0	1	0	0	1
Bit Position	15 (MSB)	14	13	12	11	10	9	8

Bit	HOLD	NMIES	NMI	TMSEL	CNTCL	SSEL	IS.1	IS.0
Reset Value	0	0	0	0	0	0	0	0
Bit Position	7	6	5	4	3	2	1	0 (LSB)

PWD : Password Byte, as described above.

HOLD : When set, the WDT is halted

NMIES : Non-Maskable Interrupt Edge Select
NMIES=0: If enabled, the NMI occurs on rising clock edge
NMIES=1: If enabled, the NMI occurs on falling clock edge

NMI : Function Select for the RST/NMI pin
NMI=0: RST/NMI performs reset functions
NMI=1: RST/NMI acts as a non-maskable interrupt

TMSEL : Watchdog Timer Mode Select
TMSEL=0: Watchdog mode is selected
TMSEL=1: Interval-timer mode is selected

CNTCL : Counter clear bit. Writing a 1 to this bit clears WDTCNT

SSEL : WDTCNT Source Select
SSEL=0: SMCLK sources WDTCNT
SSEL=1: ACLK sources WDTCNT

IS : Interval Select. This selects the value that the WDT count time is multiplied by.
IS=00: 2^{15}
IS=01: 2^{13}
IS=10: 2^{9}
IS=11: 2^{6}

Timer A

Timer A is a general purpose 16-bit counter and event timer, which is implemented on all three families. It is a multi-mode timer with multiple independent capture and compare registers. Timer A may be sourced by any internal clock, and can generate interrupts.

Capture and Compare Units

The Timer A and Timer B units contain some number (typically three, five or seven, depending on device) of independent capture and compare units. They operate in one of two modes, as selected by the mode bit CAP in their individual Cap/Com Control register, CCTLx.

In compare mode, the value to be compared to is loaded into the CCR register. When the timer value is equal to the value in the CCR register, an interrupt is generated. When using CCR0, the timer has a selectable mode in which the timer register resets to zero after reaching the compare value.

One of the common applications of compare mode is the Pulse Width Modulator (PWM). Description and implementation of the PWM is described in Chapter 11. The important variable here is OMOD, which resides in the Capture/Compare Control register (described below). This three-bit variable controls the nature of the output pin signal for the Cap/Com unit. The variable is described in the table below:

Table 4.1:
Compare Mode Output Signal Modes

OMOD	Output Description
000	OUTx is defined by the OUTx bit in CCTLx
001	OUTx is set when Timer=CCRx, and remains set until timer is reset
010	OUTx is toggled when Timer=CCRx, and reset when Timer=CCR0
011	OUTx is set when Timer=CCRx, and reset when Timer=CCR0
100	OUTx is toggled when Timer=CCRx
101	OUTx is reset when Timer=CCRx
110	OUTx is toggled when Timer=CCRx, and set when Timer=CCR0
111	OUTx is reset when Timer=CCRx, and set when Timer=CCR0

Generally speaking, modes 2,3,6 and 7 are used for PWM implementation, modes 1 and 5 are used for single event generation, and mode 4 is used to produce a signal that is ½ the frequency of the timer signal. I have never found a nontrivial (i.e., reasonable) use for mode 0, and am interested to hear from anyone who has.

The capture mode is used to time events. The input signal is selected by the CCIS variable in the CCTL register, with the capture edge selected by the CAPM variable located in the same register. When the proper edge is detected on the selected input line, the value in the Timer register is latched into the CCR register, providing a time base for the event. The CCIS variable can be set to supply or ground levels, so that software is able to generate events as well. This is handy for measuring how long particular algorithms require for processing.

The Capture/Compare registers are summarized later in this chapter.

Timer Operating Modes

- Mode 0: (MC=00) Stop Mode.

  The timer is stopped. The status of the timer, including the value in TAR and all control registers and flags, remain preserved in this mode.

- Mode 1: (MC=01) Up Mode.

  The timer counts up to the value in CCR0, and resets to zero. TAIFG, the general Timer A interrupt flag, is set when TAR resets to zero, if it has been enabled. CCIFG0, if enabled, is set one transition earlier, when TAR reaches the value in CCR0.

- Mode 2: (MC=10) Continuous Mode.

  This mode is similar to mode 1, except the timer automatically runs to 0xFFFF, or 65,535, and resets to zero.

- Mode 3: (MC=011) Up-Down Mode.

  The timer counts from 0 to 0xFFFF, and then counts back down to 0. This is the only operating mode in which TAR decrements.

Timer A Register Summaries

- TAR, Timer A Register
  Address: 0x0170h
  All bits are readable and writable. This register is the location of the Timer A count.

- TACTL, Timer A Control Register.
  Address: 0x0160h
  All bits are readable and writable.
  Structure:

Bit	Unused	Unused	Unused	Unused	Unused	Unused	SSEL1	SSEL0
Reset Value	0	0	0	0	0	0	0	0
Bit Position	15 (MSB)	14	13	12	11	10	9	8

Bit	ID1	ID0	MC1	MC0	Unused	CLR	TAIE	TAIFG
Reset Value	0	0	0	0	0	0	0	0
Bit Position	7	6	5	4	3	2	1	0 (LSB)

SSEL : Input Clock Select
SSEL=00: Varies by device: See data sheet
SSEL=01: ACLK
SSEL=10: SMCLK
SSEL=11: Varies by device: See data sheet

ID : Selects the value the input clock is divided by.
ID=00: 1 (Input clock is passed directly to timer)
ID=01: 2
ID=10: 4
ID=11: 8

MC : Mode Control
MC=00: Timer is stopped.
MC=01: Timer counts up to CCR0 and restarts at 0.

MC=10:	Timer counts up to 0xFFFFh and restarts at 0.
MC=11:	Timer counts up to CCR0 and back down to 0.
CLR:	Clear bit. Setting this bit clears the timer and ID bits. The clear bit automatically resets.
TAIE:	Timer A interrupt enable. If set, an interrupt is generated on timer overflow.
TAIFG:	Timer A interrupt flag. This is set when the timer resets to 0000h from any other value.

- TACCTLx, Capture/Compare Control Registers.
  Address: Vary by Cap/Compare unit: See data sheet
  Bits 9 and 3 are read-only. All other bits are readable and writable.
  Structure:

Bit	CAPM.1	CAPM.0	CCIS.1	CCIS.0	SCS	SCCI	Unused	CAP
Reset Value	0	0	0	0	0	0	0	0
Bit Position	15 (MSB)	14	13	12	11	10	9	8

Bit	OMOD.2	OMOD.1	OMOD.0	CCIE	CCI	OUT	COV	CCIFG
Reset Value	0	0	0	0	0	0	0	0
Bit Position	7	6	5	4	3	2	1	0 (LSB)

CAPM : Capture Mode

CAPM =00:	Capture Mode Disabled
CAPM =01:	Capture on rising edge
CAPM =10:	Capture on falling edge
CAPM =11:	Capture on both edges

CCIS	: Input Select CCIS =00: CCIxA is selected CCIS =01: CCIxB is selected CCIS =10: GND is selected CCIS =11: Vcc is selected
SCS	: Capture synchronization bit. If set, the capture is synchronized with the timer clock.
SCCI	: This read-only bit reflects the latched input signal. This bit is not implemented on Timer B.
CAP	: Mode select. When set, the module is in capture mode. When cleared, the module is in compare mode.
OMOD	: Output Mode Select Bits (see previous descriptions)
CCIE	: If set, capture/compare interrupt is enabled.
CCI	: The selected input signal is readable by this bit.
OUT	: This bit sets/clears the value of OUTx, when in output only mode.
COV	: Capture Overflow Bit. This bit is set if a capture occurs when an unread capture value exists. This bit is unused in compare mode.
CCIFG	: Capture/Compare interrupt flag. If CCIFG0, this flag is automatically reset. If any other CCIFG, this flag is reset when TAIV is read.

- TAIV, Timer A Interrupt Vector Register.
  Address: 0x012E
  Structure:

Bit	Unused	Unused	Unused	Unused	Unused	Unused	Unused	Unused
Reset Value	0	0	0	0	0	0	0	0
Bit Position	15 (MSB)	14	13	12	11	10	9	8

Bit	Unused	Unused	Unused	Unused	IV.2	IV.1	IV.0	Unused
Reset Value	0	0	0	0	0	0	0	0
Bit Position	7	6	5	4	3	2	1	0 (LSB)

This read only register identifies which capture/compare module generated an interrupt. It is interesting to note that IV is defined as a three-bit number, but only even values are valid, so IV.0 is always cleared. Consult the TI documentation for interpretation of IV values.

Timer B

Timer B is a second independent timer circuit, nearly identical to Timer A. The only difference is the SCCI bit in the capture/compare modules, which is not implemented in these timers. Otherwise, the register descriptions for Timer A are valid for Timer B. TBCTL is located at address 0x0180, TBR is at 0x0190, and TBIV is at 0x011E.

Debugging Clock Difficulties

I have, over the course of time, developed a three-step process for debugging clock problems. While it will not find all problems all the time, it has proved useful in most situations. It has been my experience that the vast majority of clock problems stem from one of three areas: hardware, unexpected interrupt generation, or misconfiguration of internal clocks.

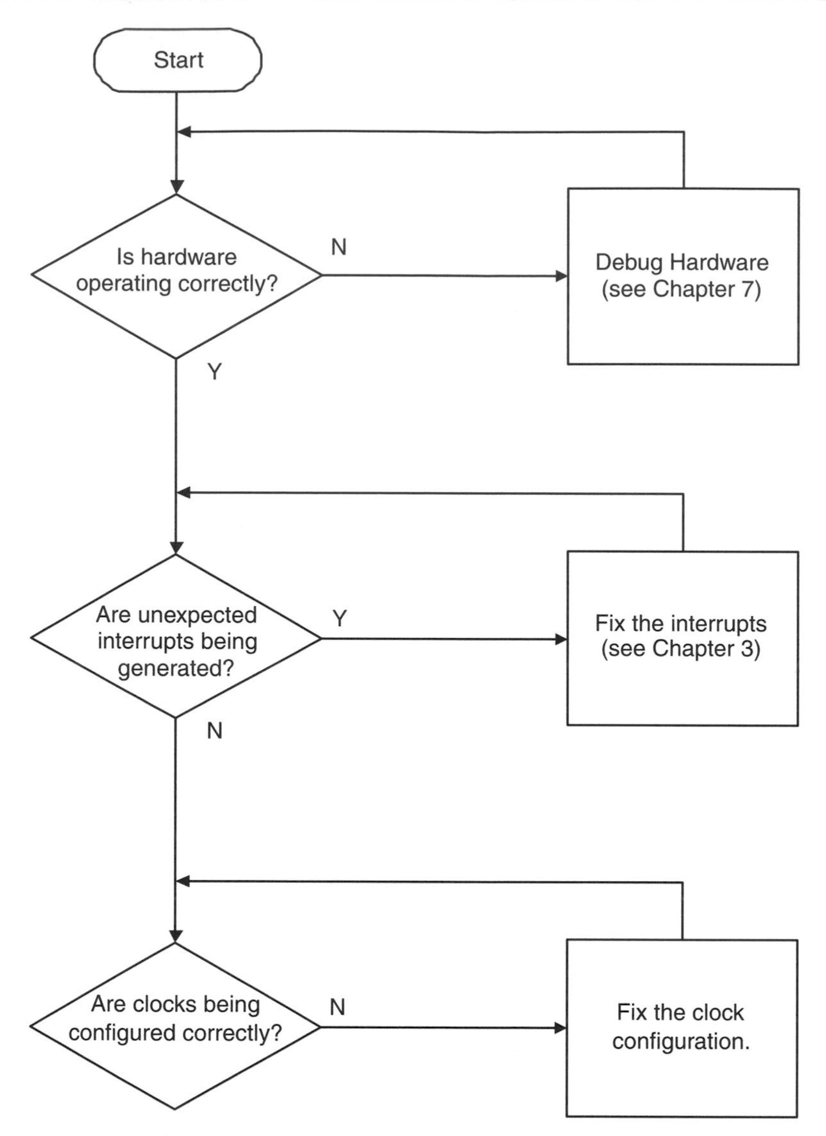

Figure 4.8: Clock Debugging Flowchart

- *Hardware*. This is a particularly common area of difficulty. Hardware debugging is discussed further in Chapter 7, but a few clock-specific suggestions are offered here. As with any hardware debugging, your best friend is a good digital oscilloscope. Use the scope to observe the crystal outputs, at the high pin, throughout the operation. Look at the frequency and level, and look for dropouts and glitches. (I will sometimes use freeze spray while running this test, to observe crystal accuracy over temperature.) Look at the supply, ground, and signal lines for electrical transients.
- *Unexpected Interrupt Generation*. Although unexpected interrupts are seldom the direct cause of this type of problem, they often alter the timing of your design enough to appear as clock errors. They are detectible with the use of while loops defined in Chapter 3, or with judicious use of breakpoints.
- *Clock Configuration*. Make certain your internal clocks are configured properly. With the MSP430, this is very easy to identify. On most devices, internal clocks such as MCLK, ACLK, and SMCLK can be observed externally, by using the function select on the appropriate port (see the device datasheet and Chapter 5 for proper pin selection and configuration).

As mentioned above, this is by no means comprehensive. You probably noticed that the flowchart in Figure 4.8 has no end state. This is because you will not always have solved your problem by following the steps it describes. However, these three checks have almost always pointed me in the right direction.

Crystal Accuracy

Any timed event or process cannot be any more accurate than the clock source from which it derives. Because of this, oscillator accuracy is a common, fundamental source of difficulty when designing systems. It is important to understand how accurate your oscillators will be.

The first rule is, if your design needs any reasonable degree of timing precision, use a crystal. The DCO is a simple RC-type oscillator, and the variation over temperature is far too great to perform anything resembling reasonably accurate timing. A decent crystal will be on the order of ten to 100 times more accurate over temperature and voltage than the DCO.

Once you have settled on the use of a crystal, make certain that you understand its specifications and limitations. The error values are typically specified in parts per million (ppm), and there are three different components to be aware of:

- *Finishing Tolerance*. This is the maximum inaccuracy that results from inconsistencies in the manufacturing process. It is the initial offset from the ideal frequency at the nominal temperature. It is typically on the order of +/- 30 ppm, and is a limitation over the entire temperature range and lifetime of the crystal.
- *Aging Tolerance*. This is the change of accuracy over time. It is typically just a few ppm/year, and is tends to be more pronounced in the first few years (e.g. +/-3 ppm the first year and +/- 1 ppm in succeeding years). The effect is additive over time.
- *Temperature Tolerance*. This is usually the most significant component of error, and varies widely from one crystal to another. It is often specified in either ppm over a specified temperature range, or in ppm/°C. Some manufacturers will simply give a graph or equation. (An example graph is given in Figure 4.9.)

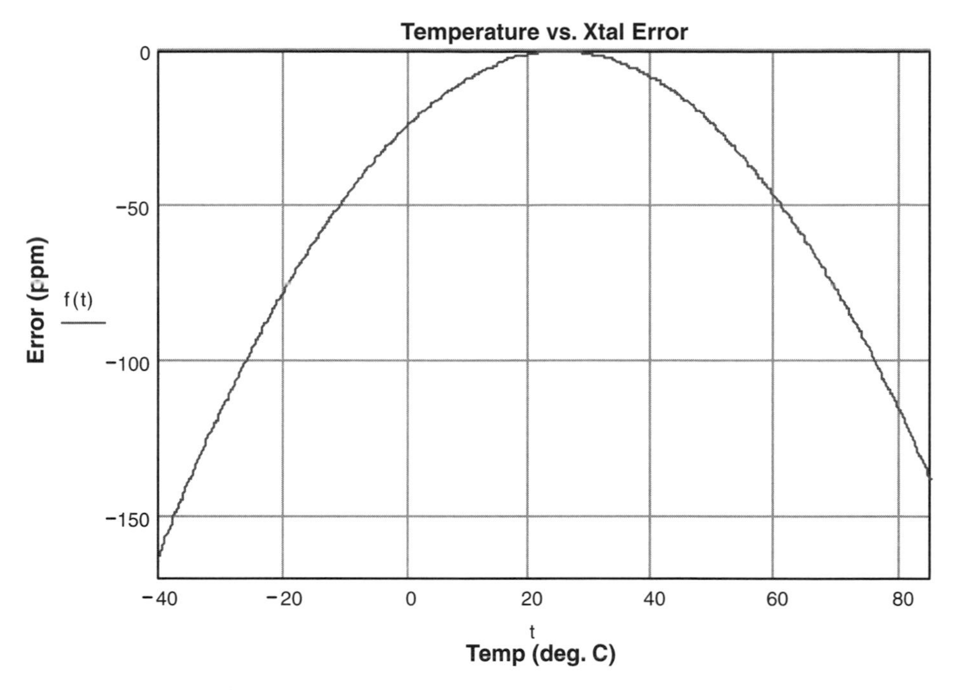

Figure 4.9: Sample Crystal Error Curve

Often, crystals are selected based on temperature tolerance only. Low-end 32 kHz crystals tend to run around 10 ppm/°C, so the aging and finishing numbers are lost in the noise. Higher end, more expensive crystals tend to also have tight aging and finishing tolerance, so temperature has still the biggest impact.

CHAPTER 5

Input and Output

Every system you will ever design has one thing in common: use of input and/or output. (Some fully deterministic designs will not require inputs, but I cannot imagine any useful design that does not require output.) All I/Os on MSP430 devices are eight bits wide and controlled with memory-mapped registers. These ports, which are fundamentally the same across all device families, can be divided into two categories: interruptible (meaning that interrupts can be generated via these ports) and non-interruptible. In all devices, ports 1 and 2 are interruptible, and the higher numbered ports are not. '3xx devices have a port 0, which is also interruptible (there is no port 0 in '1xx and '4xx devices).

Along with basic I/O functions, the port pins can be individually configured as special function I/Os, such as USARTs, Comparator signals, and ADCs. The number of ports and available functions vary by part, so consult your datasheet.

Non-Interruptible I/O

I/O ports 3,4,5 and 6 are non-interruptible data ports. These ports are not implemented on all devices.

Use of non-interruptible I/O is simple and straightforward. Each bit is individually controllable, so inputs, outputs, and dedicated function I/O can be mixed in a single port. Port pins are controlled by four byte-addressable registers: direction, input, output, and function select.

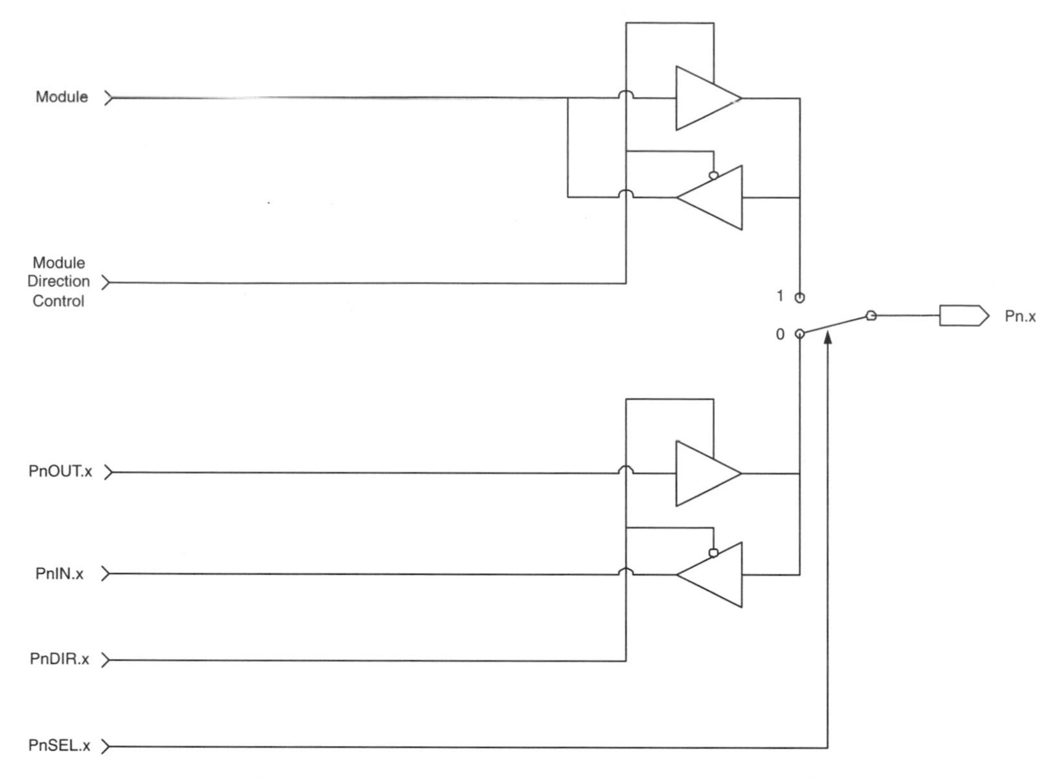

Figure 5.1: Non-Interruptible I/O Pin Logic

- *Direction Registers.* These read/write registers control the signal direction for port pins. When a bit in the direction register is set, the corresponding port pin is set as an output, and when the bit in the direction register is cleared, the port pin is set as an input. The direction registers need to be configured properly if the port pin is selected as a general purpose I/O or as a special function I/O. Direction registers are cleared on reset.
- *Input Registers.* These are read-only registers, which reflect the input value on the port.
- *Output Registers.* These registers are used to write to output ports, and can be read as well. When reading these registers, they will reflect the last value written to them. However, if the port is configured as an input, the output register will be in an indeterminate state. It will not necessarily reflect the input value on the associated pin.

- *Function Select Registers*. These read/write registers determine the use of the individual pins on the I/O port. When the bit in the select register is set, the port pin is set as a function I/O, and when the bit in the direction register is cleared, the port pin is set as general purpose I/O.

Interruptible I/O

Ports 1 and 2 (and port 0 on '3xx devices) are interruptible ports. They contain all of the same control registers as non-interruptible ports (described in the previous section), along with three other byte-addressable registers: interrupt enable, interrupt edge select, and interrupt flags.

- *Interrupt Enable*. This read-write register enables interrupts on individual pins. Interrupts on the pins are enabled when their corresponding bits in this register are set. This register is cleared on reset.

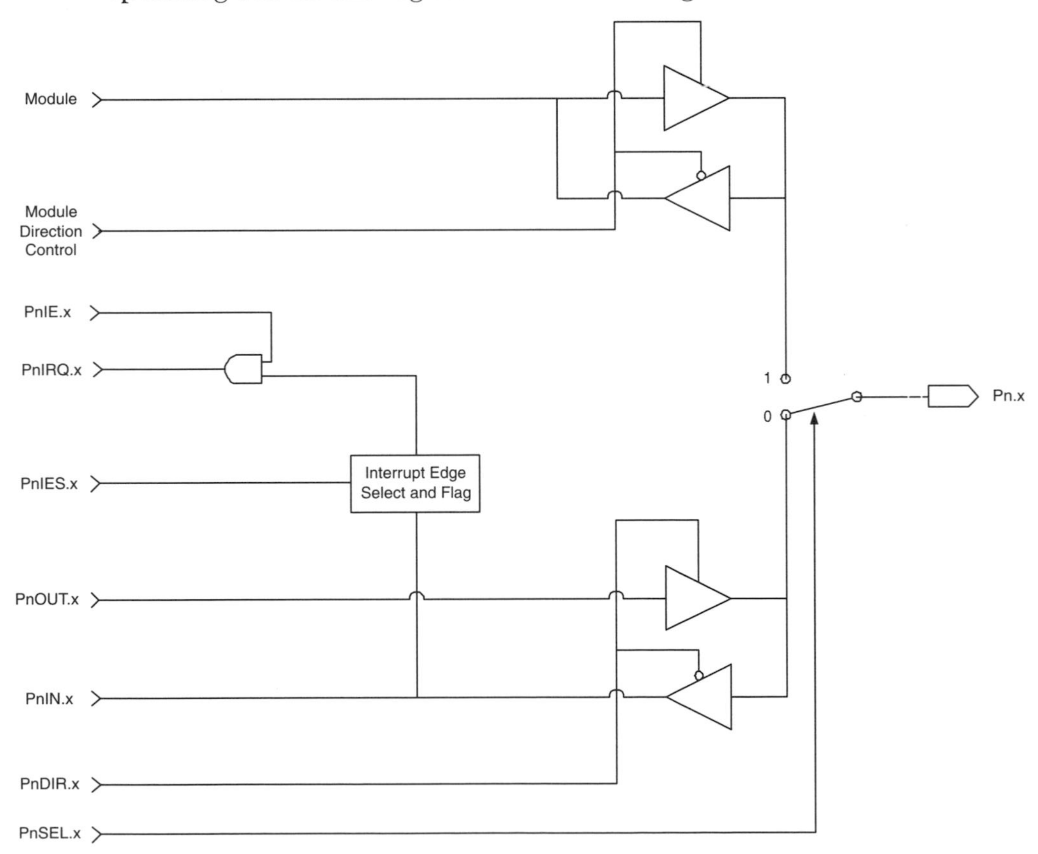

Figure 5-2: Interruptible I/O Pin Logic

- *Interrupt Edge Select*. This read-write register selects the transition on which an interrupt occurs. If set, an interrupt occurs on a high-to-low transition on the corresponding pin. If cleared, an interrupt occurs on a low-to-high transition on the corresponding pin.
- *Interrupt Flags*. The corresponding bit in this read-write register is set automatically when an interrupt is generated. This register can be written to, and will generate an interrupt when a high level is written. When an interrupt occurs, this flag needs to be cleared before the reti instruction, or the same interrupt will call the ISR a second time.

Table 5.1: Port Registers

	Port 0 ('3xx only)	Port 1	Port 2	Port 3	Port 4	Port 5	Port 6
Input	P0IN 010h	P1IN 020h	P2IN 028h	P3IN 018h	P4IN 01Ch	P5IN 030h	P6IN 034h
Output	P0OUT 011h	P1OUT 021h	P2OUT 029h	P3OUT 019h	P4OUT 01Dh	P5OUT 031h	P6OUT 035h
Direction	P0DIR 012h	P1DIR 022h	P2DIR 02Ah	P3DIR 01Ah	P4DIR 01Eh	P5DIR 032h	P6DIR 036h
Interrupt Flags	P0IFG 013h	P1IFG 023h	P2IFG 02Bh	Not Implemented	Not Implemented	Not Implemented	Not Implemented
Interrupt Edge Sel	P0IES 014h	P1IES 024h	P2IES 02Ch	Not Implemented	Not Implemented	Not Implemented	Not Implemented
Interrupt Enable	P0IE 015h	P1IE 025h	P2IE 02Dh	Not Implemented	Not Implemented	Not Implemented	Not Implemented
Function Select	Not Implemented	P1SEL 026h	P2SEL 02Eh	P3SEL 01Bh	P4SEL 01Fh	P5SEL 033h	P6SEL 037h

Using I/O

- All port registers can be changed in software. This allows software to turn interrupts on and off, and to use a single pin for both input and output.
- When port pins are configured as inputs, they also function as high-impedance inputs. When being used as a high-impedance input, no pulling resistor is necessary, as long as any externally applied voltage is at the ground or supply rail. When configured this way, the leakage current is typically on the order of tens of nanoamperes.

- Interrupts on this device are edge triggered, and very susceptible to noise. A bit of filtering on interrupt lines will go a long way in noisy designs. (This is further described in Chapter 7, *Hardware Considerations*.)

JTAG

Along with the banks of general-purpose I/O available on the MSP430 devices, there is, on many of the devices (particularly Flash parts) a JTAG interface port. This interface implements IEEE STD1149.1 compliant access to the CPU, peripherals, and internal busses. It is the primary method for erasing, programming, and checking flash memory devices, and is a very powerful tool for running test routines.

Texas Instruments has available, at their website, an exceptional application report on JTAG and its use in the MSP430 (App. Rept. SLAA149), which I will not try to summarize here. Among the handiest of features described in this paper is the command to set the PC to an arbitrary value. This feature makes it possible to have test code, built into the design, that is completely inaccessible from the main code set. The PC is set to the starting address of that code, via the JTAG, and the tests are performed. This is useful for things like bench and production code, that never needs to see the light of day once the product is sold, shipped, or deployed.

Another similar feature is the ability of the JTAG to write directly to registers, completely bypassing the CPU. This allows specialized tests which exercise the surrounding hardware, by reading and writing the Port control pins, without running firmware. This can be useful if your design lacks the code space necessary to write these one-time tests.

CHAPTER 6

On-Chip Peripherals

One of the strong features of the MSP430 is the choice of available peripherals. This chapter overviews use of several of the most common and frequently used blocks. It has been my experience that use of these produces the bulk of programming and configuration errors. Because of this, when relying on these peripherals, it is important to review thoroughly the Users Guide and pertinent application notes, and even to introduce yourself to one of the TI application engineers.

Hardware Multiplier

Some of the larger '430 devices include a hardware multiplier. It is, however, a bit different from the multipliers in many similar scale controllers. Rather than being a fixed function of the ALU, which is common, the '430 hardware multiplier is implemented as a memory-mapped peripheral device. It is pretty straightforward to use. Write operands to two registers, in order, wait a few clock cycles, and the result is magically located in the result registers.

The module performs both multiply and multiply-and-accumulate functions, and can perform either as signed or unsigned multiply. These options are selected when loading the first operand. There are four different operand 1 registers, one for each multiplication type. The second operand register is universal, and the multiplier activates when a value is loaded into this register. The result is found in two registers, a Result LO register and a Result HI register.

There is also a Sum Extend register, which serves several purposes. In unsigned multiply mode, it remains unused. In signed multiply mode, it reflects the sign of the result (0x0000 for positive, 0xFFFF for negative). In MAC mode, it is normally zero, but if the result exceeds 0xFFFF FFFF, the multiplier writes a 0x0001 to the register. In signed MAC mode, the behavior is similar to the signed multiply, except the register holds a value of 0x0000 if the result is 0x7FFF FFFF or less (positive), and 0xFFFF if the result is 0x8000 0000 or greater (negative, 2's complement).

Table 6.1: Hardware Multiplier Registers

Register	Address	Read/Write
Operand 1: Unsigned Multiply	0x0130	Read/Write
Operand 1: Signed Multiply	0x0132	Read/Write
Operand 1: Unsigned Multiply and Accumulate	0x0134	Read/Write
Operand 1: Signed Multiply and Accumulate	0x0136	Read/Write
Operand 2	0x0138	Read/Write
Result LO	0x013A	Read/Write
Result HI	0x013C	Read/Write
Sum Extend	0x013E	Read Only

Sources of Error with the Hardware Multiplier

There are several common areas of difficulty that the user needs to be aware of when using the hardware multiplier.

- *Existence*. Don't laugh, this actually happens. Take the example of the MSP430F149. This is a very common device to perform initial development and proof-of-concept work on, because TI offers a low-cost FET development tool based around the device, and it is pin-for-pin compatible with '13x devices. I have, in the past, done initial development on the '149, and then switched to the '133 or '135 for final product. Well, the '13x devices don't have the multiplier, and my code was reading and writing phantom registers. It only took a few minutes to troubleshoot, but was, needless to say, embarrassing. (It is also embarrassing to write here, but we all make the occasional bonehead mistake, right?)

- *Premature access to the result.* The multiplication process is not immediate, it takes a few clock cycles. The Users Guide says that the line is between autoincrement and indexed modes (with autoincrement, you need a statement between loading of operand and retrieval of result, with indexed, you do not). I will take their word for it. I have always thrown a NOP in there, just to be safe.
- *Overflow.* The multiplier does not explicitly report overflow (via the overflow flag) in Multiply and Accumulate mode. Rather, it changes the Sum Extend register. The application needs to be prepared for this.
- *Interrupts.* If your code uses the multiplier in the main loop and one of the ISRs, there exists an error condition when the main loop loads one or both operands, and the interrupt occurs before the main loop retrieves the result. The ISR then uses the multiplier, and the main loop retrieves an incorrect value after return form interrupt. There are various ways to handle this, none of them being very good. If you need to perform multiplication in ISRs and the main loop, the best solution is to write a 16 × 16 multiply function, for use in the main loop, and restrict hardware multiplier use to the ISRs (or vice versa).

Comparator

Some MSP430 devices offer an on-board analog comparator, which is very basic and easy to use. These comparators are configurable such that either input or output (or both) may be mapped to external pins. It has an internal reference voltage generator, which can be turned off to conserve current, or set to ¼ or ½ of Vcc, the device supply voltage. The comparator can generate interrupts on either the rising or falling edge of the output.

The functionality is essentially described by the function registers. The biggest source of potential error is that of output oscillation. When the levels at the inputs are very close to each other, the output tends to oscillate. To address this problem, the device has an RC-type filter on the output, which may be switched in or out. This filter does not completely solve the oscillation problem. If your application still oscillates after switching this filter in, you will probably need to use an external precision comparator instead, especially if the comparator is driving an interrupt.

Comparator Control Registers

- CACTL1, Comparator control register #1.
  Address: 0x0059h
  All bits are read/write.
  Structure:

Bit	CAEX	RSEL	REF.1	REF.0	CAON	CAIES	CAIE	CAIFG
Reset Value	0	0	0	0	0	0	0	0

CAEX (MSB) : Comparator Exchange.
When set, this bit swaps the inputs of the comparator.

RSEL : Reference Select.
This bit selects where the internal reference is applied. (The CAEX bit swaps this)
0=Reference is applied to the + terminal.
1=Reference is applied to the - terminal

REF : Sets internal reference.

REF=0: Off
REF=1: Reference = Vcc/4
REF=2: Reference = Vcc/2
REF=3: Diode reference is selected. See datasheet for details on this option.

CAON : Comparator On.
When set, the comparator is on. When cleared, it is off.

CAIES	: Comparator Interrupt Edge Select. 0= Interrupt occurs on rising edge of Comparator output. 1= Interrupt occurs on falling edge of Comparator output.
CAIE	: Comparator Interrupt enable. Interrupt is enabled when bit is set.
CAIFG (LSB)	: Comparator Interrupt Flag.

- CACTL2, Comparator control register #2.
  Address: 0x005Ah
  All bits except CAOUT are read/write. CAOUT is read only.
  Structure:

Bit	CCTL2.7	CCTL2.6	CCTL2.5	CCTL2.4	P2CA1	P2CA0	CAF	CAOUT
Reset Value	0	0	0	0	0	0	0	0

CCTL2 (MSB)	: These bits vary by device. See the datasheet for details.
P2CA1	: Pin to CA1. If set, the CA1 is mapped to the external pin.
P2CA0	: Pin to CA0. If set, the CA0 is mapped to the external pin.
CAF	: Comparator Output Filter 0= Comparator Output Filter is bypassed. 1= Comparator Output Filter is used.
CAIE	: Comparator Interrupt enable. Interrupt is enabled when bit is set.
CAOUT (LSB)	: Comparator output.

- CAPD, Comparator Port Disable.
  Address: 0x005Ah
  All bits are read/write.
  Structure:

Bit	CAPD.7	CAPD.6	CAPD.5	CAPD.4	CAPD.3	CAPD.2	CAPD.1	CAPD.0
Reset Value	0	0	0	0	0	0	0	0

CAPD.x (All): These pins enable (0) or disable (1) the input buffer pins for up to eight I/O. Not all eight are implemented on all devices.

Comparator Setup Examples

Code Example 6.1: Various Comparator Setup Schemes

```
/*  External inputs and output, no interrupt use, no output filter, 'F149
    pin configuration   */
   {
    P2SEL |= 0x1C;                     //Set up I/O direction register
    CACTL1 = CAON;                     //Turn on Comparator, no
                                       //internal reference use
    CACTL2 = P2CA0 + P2CA1;            //Use external inputs
    CAPD = 0xFC;                       //Enable both CAPD input
                                       //buffers
    }

/*  One external input with internal reference, interrupt driven, output
    filter, 'F149 pin configuration   */
   {
    P2SEL |= 0x10;                     //Set up I/O direction register
    CACTL1 = CAIE + CAREF_2 + CAON;  //Turn on Comparator,
                                        //internalVcc/2 reference
    CACTL2 = CAF + P2CA1;              //Use external signal  for –
                                       //comparator input
```

```
    CAPD = 0xFD;                        //Enable CAPD1 input buffer
    }

    interrupt [0x16] void ComparatorA_Interrupt(void)
    {
    /*  Process interrupt here      */
    }

/*  One external input with internal reference, no interrupt, single decision,
    output filter, 'F149 pin configuration   */
   {
    P2SEL |= 0x10;                      //Set up I/O direction register
    CACTL1 = CAREF_1 + CAON;            //Turn on Comparator, internal
                                        //Vcc/4 reference
    CACTL2 = CAF + P2CA1;               //Use external signal for –
                                        //comparator input
    CAPD = 0xFD;                        //Enable CAPD1 input buffer
    for (index=0;index<15;index++) {}   //Loop delay allows filter output
                                        //to stabilize
                                        //the length will depend on
                                        //clock speed

    if (CACTL2 & CAOUT)
       {
       /* Process Positive Case */
       }
    else
       {
       /* Process Negative Case */
       }
    }
```

Analog-to-Digital Converters

Most MSP430 devices offer a precision analog-to-digital converter. There are various flavors of converter available. We will describe the most common, the ADC12. If your device offers a different unit, it is very similar to the 12-bit version described here, and the relevant differences and detail are available in the user's guide.

It is important to note that the ADC is by far the most complex of the peripherals available for the MSP430. My intent here is to cover the high points of its use, but my description is by no means complete. If you intend to use the ADC, first become intimate with the contents of the user's guide. I only hope to give the reader a feel for the capabilities and limitations of the peripheral.

The ADC12 is a single 12-bit analog-to-digital converter, with a built-in sample-and-hold circuit. The front end consists of a multiplexer circuit, allowing the developer to select one of eight external pins, or one of four internal sources, for the signal to convert. The most interesting of these is an internal temperature diode, which allows the ADC to provide a rough idea of operating temperature. The temperature diode varies by device, and the details are in the datasheet. It is considerably less accurate than a true external temperature sensor, but it can be a useful low-cost alternative in some applications.

There are four conversion modes, reflecting the permutations of single and multiple conversions and one-time and repeated conversions. Specifically, they are:

- Single channel one-shot. This is the true single conversion, with the result being stored in one of the ADCMEM registers.
- Single channel repeated. As the name would suggest, this mode repetitively performs conversion until stopped, storing the result in the same ADCMEM register. In this mode, the typical method is to loop process when the BUSY flag clears.

- Multiple channels, single sequence. In this method, the ADC performs multiple conversions, looping through a specified number of ADCMEM registers one time.
- Multiple channels repeated. This is identical to the previous case, except the series of conversions is repeated until stopped.

One of the nice things about these modes is that they are all fire-and-forget, meaning that you can write the code so that the process is initialized, and the code can perform other processing while the conversion is underway. The biggest limitation is that, in the repeated modes, the software needs to be ready to read the ADCMEM registers before they are rewritten, or an interrupt will be generated.

Timing for the ADC is performed by the conversion clock. The conversion clock may be sourced by any of the clocks from the Basic Clock Module, or by a fixed RC oscillator, which is a dedicated portion of the ADC. This oscillator is very similar to the DCO, with the same limits of accuracy. It is however, a fixed speed oscillator, running around the max DCO frequency, which is just shy of 5 MHz (typical). This is nice to have, so that the ADC can operate while the CPU is in sleep mode. The timer must be initialized in the ADC12CTL1 register, and the DIV value and clock source must be selected such that the conversion frequency meets the datasheet spec. The conversion takes thirteen cycles of the conversion clock source.

ADC12 Control Registers

- ADC12CTL0, ADC control register #0.
  Address: 0x01A0h
  All bits are read/write. Bits 15 through 4 may only be edited when ENC=0.

Structure:

Bit	SHT1.3	SHT1.2	SHT1.1	SHT1.0	SHT0.3	SHT0.2	SHT0.1	SHT0.0
Reset Value	0	0	0	0	0	0	0	0
Bit Position	15 (MSB)	14	13	12	11	10	9	8

Bit	MSC	2_5V	REFON	ADCON	ADCOIE	ADCTIE	ENC	ADC_SC
Reset Value	0	0	0	0	0	0	0	0
Bit Position	7	6	5	4	3	2	1	0 (LSB)

SHT1/SHT0 : Sample and Hold Time

SHT1 determines the sample and hold time for ADC0 through ADC7, and SHT1 determines sample and hold time for ADC8 through ADC 15. The sample and hold time is 4x(ADC_Clock_Time)x(n). See table 6.2 for mapping of SHT values to n values.

MSC : Multiple Sample and Convert:

0: The sampling timer requires a rising edge on SHI.

1: SHI triggers the first sample, all subsequent samples are automatic.

2_5V : Internal Reference Voltage level

0: 1.5 VDC

1: 2.5 VDC

REFON : Internal Reference Voltage enable

0: Internal Reference is off

1: Internal Reference is on

ADCON : Analog to Digital Converter ON

0: ADC is off

1: ADC is on

ADCOIE : ADC Overflow Interrupt Enable
If set, an interrupt is generated on overflow. ADC overflow is defined as the condition that occurs when a result is written and the previous result has not been read.

ADCTIE : ADC Timer Overflow Interrupt Enable
If set, an interrupt is generated on ADC timer overflow. ADC timer overflow is defined as the condition that occurs when a conversion is attempted before the previous conversion is attempted.

ENC : Enable Conversion
If cleared, no conversion is possible.
If set, a conversion is possible.
The ADCMCTL registers cannot be edited unless this bit Is cleared. The typical process is to set the ADC up, and then bring this bit high.

ADC_SC : ADC Sample and Convert
When set, the conversion process automatically begins (if ENC=1). This bit is automatically reset when the conversion is complete.

Table 6.2: n Values for SHT variable

SHT Value	n Value
0	1
1	2
2	4
3	8
4	16
5	24
6	32

7	48
8	64
9	96
10	128
11	192
12	256
13	256
14	256
15	256

- ADC12CTL1, ADC control register #1.
  Address: 0x01A2h
  All bits except 0 (BUSY) are read/write. Bits 15 through 3 may only be edited when ENC=0.
  Structure:

Bit	CSAdd.3	CSAdd.2	CSAdd.1	CSAdd.0	SHS.1	SHS.0	SHP	ISSH
Reset Value	0	0	0	0	0	0	0	0
Bit Position	15 (MSB)	14	13	12	11	10	9	8

Bit	DIV.2	DIV.1	DIV.0	SSEL.1	SSEL.0	CONS.1	CONS.0	BUSY
Reset Value	0	0	0	0	0	0	0	0
Bit Position	7	6	5	4	3	2	1	0 (LSB)

CSAdd : Conversion Start Address
Selects which ADC12MEM register is used for the first conversion in a sequence.

SHS : Source Select for the Sample-Input Signal.
00: Control Bit ADC12SC
01: Timer_A.OUT.1
10: Timer_B.OUT.0
11: Timer_B.OUT.1

SHP : Sampling Signal Select

- 0: SAMPCOM is sourced from the sample-input signal.
- 1: SAMPCOM is sourced from the rising edge of the sampling timer.

ISSH : Invert Sample-Input Signal

- 0: The Sample-Input Signal is normally defined.
- 1: The Sample-Input Signal is inverted.

DIV : Clock Division Rate

The clock selected by SSEL is divided by DIV+1

SSEL : Clock Source Select

- 00: ADC Internal Oscillator
- 01: ACLK
- 10: MCLK
- 11: SMCLK

CONS : Conversion Mode Select

- 00: Single-Channel, Single Conversion
- 01: Single Sequence of Channels
- 10: Repeat in a single channel until CONS is changed
- 11: Repeat in a sequence of channels until CONS is changed.

BUSY : Busy Flag (read-only)

This read-only flag is set while a conversion is underway.

- ADC12MCTLn, individual ADC control registers 0 through 15.
  Addresses: 0x0080h through 0x008F
  All bits are read/write. All bits may only be edited when ENC=0.
  Structure:

Bit	EOS	SREF.2	SREF.1	SREF.0	INCH.3	INCH.2	INCH.1	INCH.0
Reset Value	0	0	0	0	0	0	0	0
Bit Position	7	6	5	4	3	2	1	0 (LSB)

EOS : End Of Sequence
This bit indicates when a sequence of conversions is complete.

SREF : Reference Voltage Select
000: V+ is AVcc, V- is AVss
001: V+ is Vref+, V- is AVss
010: V+ is Veref+, V- is AVss
011: V+ is Veref+, V- is AVss
100: V+ is AVcc, V- is (Vref-)/(Veref-)
101: V+ is Vref+, V- is (Vref-)/(Veref-)
110: V+ is Veref+, V- is (Vref-)/(Veref-)
111: V+ is Veref+, V- is (Vref-)/(Veref-)

INCH : Input Channel Select
0-7: External a0 through a7
8: Veref+
9: (Vref-)/(Veref-)
10: Internal Temperature Diode
11-15: (Avcc-Avss)/2

- ADC12IFG, ADC Interrupt Flags
  Address: 0x01A4h
  All bits are read/write.

- ADC12IEN, ADC Interrupt Enables
  Address: 0x01A6h
  All bits are read/write.

These are standard interrupt flags and enable bits, with bit 15 of each field corresponding to ADC12MEM15, etc. If the ADCMEM registers are accessed, their corresponding interrupt flags are automatically reset.

- ADC12MEM, Conversion Memory Registers 0-15
  Addresses: 0x0140h-0x0156h
  All significant bits (11-0) are read/write. Bits 15-12 read as 0.
  These 16 registers hold the result of conversions in the bottom 12 bits.

An ADC Example

As described previously, the ADC units on many devices include a temperature sensor. This nice little add-on feature is easy to use and is pretty accurate (typically within a degree or so). Code listing 6.1 provides a C-language module for reading temperature. Several considerations are:

- This code is written assuming 3.35 mV/°C, as is the case with the '149 device. If you are using a different device, check your datasheet for the correct value.
- The conversions from ADC reading to °C and from there to °F are designed to avoid use of floating-point mathematics. They will result in some additional error. However, the total error from both conversion and mathematics will still keep you within a degree or two. If your application requires precise measurement of temperature, consider an external sensor.
- The variable sizing of this function is a bit on the inefficient side. It is overkill to report a value between –50 and +200 using a signed long, but the conversion to temperature requires more than 15 bits of precision. In fact, if we turn the function into a signed int, it should automatically return the bottom 16 bits of the Temperature variable, which will still give us the correct value. However, various experiences doing things like this have left me distrustful of microcontroller compilers and their ability to always do what they I think they ought to.

CodeListing 6.2: Temperature Determination using the ADC-12

```
signed long  Determine_Temp (unsigned char BooleanFlag)
//  This function returns device temperature, based on an ADC
    conversion
//  The passed parameter acts as a flag. If it is zero, the temperature is
//  returned in degrees Fahrenheit. If it is non-zero, it is returned in
//  degrees Celsius.
{
signed long Temperature=25;      Define and initialize working variable

ADC12CTL0 = SHT0_6 + SHT1_6 + REFON + ADC12ON;
// Initialize control register 0. Sets up conversion clock, internal reference
ADC12CTL1 = SHP;                 //Conversion sample timer signal is
                                 //sourced from the sampling timer
ADC12MCTL0 = INCH_10 + SREF_1;   //Use internal reference with
                                 //respect to ground, Temp sensor
                                 //is on input channel 10.
ADC12CTL0 |= ADC12SC + ENC;      //Enable and start conversion
while (ADC12CTL1 & 0x01);        //Hang in loop until conversion
                                 //completes.
Temperature = ADC12MEM0 & 0x00000FFF;
Temperature  *= 845;             //These steps convert the ADC reading
                                 //to degrees Celsius
Temperature  >> = 13;
Temperature  -= 278;
if (BooleanFlag) return Temperature;   //Return temperature in degrees
                                       //Celsius, if required
Temperature *=461;
Temperature >.=8;                //Otherwise, convert to Fahrenheit and
                                 //return
Temperature +=32;
return Temperature;

}//End Determine Temperature Function
```

LCD Driver

The '3xx and '4xx families include liquid crystal display (LCD) driver, capable of supporting multiplex rates up to 4. The differences between the '3xx and '4xx drivers are relatively insignificant, consisting primarily of addresses and number of segments supported. Specific information in this chapter is given for the '4xx series devices, but the concepts carry over to the '3xx devices, as well.

Throughout the following discussion, I have assumed that you, the reader, have some working knowledge of the display that will be interfaced to the '430 device. If you are still a bit fuzzy on the concepts of multiplexing or 7 segment display layout, these topics are covered pretty well in the Users Guides, and probably on the display datasheet as well.

There are two significant analog components that are central to the LCD driver. The first is the timing generator. This is sourced from Basic Timer 1 (see Chapter 4), and needs to be configured according to the equation:

$$\text{LCD Freq.} = 2 \times (\text{Multiplex Rate}) \times (\text{Framing Frequency}),$$

where the framing frequency is typically a range given on the LCD datasheet. This is the only configuration item required by the timing generator.

The second analog component is the voltage generator, which produces appropriate internal drive voltages from (up to) 4 externally supplied voltage levels. These external signals are identified on datasheets as R33, R23, R13, and R03. Their intended use is reflected in the naming convention, as R33 is to be connected to full-scale voltage, R23 is 2/3 of full scale, R13 is 1/3 of full scale, and R03 is ground. There is a special case, in multiplexer mode 2, where R23 is unused and R13 should be configured as R33/2. In some devices, R33 and R03 are not implemented as separate pins. Rather, they are internally tied to supply and ground. Often these voltages are established with a resistor-divider network, to preserve these ratios. Use of these pins is determined by the multiplexer mode, and is identified in the following table:

Analog Inputs Used by Multiplexer Mode

Multiplexer Mode	R33	R23	R13	R03
Single (Static) Mode	X			X
2 Mux	X		X	X
3 Mux	X	X	X	X
4 Mux	X	X	X	X

There is a single control register for the LCD driver unit

- LCDCTL, LCD control register.
  Address: 0x0030h (in '4xx devices)
  All bits are read/write.
  Structure:

Bit	LCDM.7	LCDM.6	LCDM.5	LCDM.4	LCDM.3	LCDM.2	LCDM.1	LCDM.0
Reset Value	0	0	0	0	0	0	0	0

LCDM.7-LCDM.5 : Port Select. These signals determine how many I/O pins are selected for use by the LCD driver. See accompanying table.

LCDM.4-LCDM.3 : Multiplexer Mode
00 = Static mode
01 = 2 Multiplexer Mode
10 = 3 Multiplexer Mode
11 = 4 Multiplexer Mode

LCDM.2 : Segment Select. When LCDM=0, all segments are shut off. This is a handy feature for making the LCD blink.

LCDM.1 : Unused

LCDM.0 : LCD Driver enable. When this bit is cleared, the timing generator is off, and all lines are brought low.

LCD Driver Port Select

LCDM.7	LCDM.6	LCDM.5	Segment Pins Used
0	0	0	No Segment Pins Selected
0	0	1	S0 – S15
0	1	0	S0 – S19
0	1	1	S0 – S23
1	0	0	S0 – S27
1	0	1	S0 – S31
1	1	0	S0 – S35
1	1	1	S0 – S39

Now that the hardware and control register are set up, we need to code the thing. The firmware to the driver simply writes segment definitions to predefined memory locations. In the '4xx family, these locations begin at 0x0091 and run through 0x00A0 (inclusive). The values written to these locations depend heavily on the multiplexing and external connections to these locations. Some helpful (and not so helpful) hints:

- Triple-check all connections. This is, generally speaking, a good idea. However, if a GPIO is connected wrong, it is usually intuitively obvious which one needs to be fixed. With the LCD, there are an awful lot of connections that are more difficult to troubleshoot on the fly, especially when multiplexing 3 or 4 signals.
- Be prepared to take some time to muddle through it. This peripheral is, at best, poorly documented by TI. (Admittedly, the documentation here is not much better. None of the examples I had were any better than the ones in the User's Guide, so I will just refer you to those.)
- Make certain that you completely understand the multiplexing scheme used by the LCD. I once burned an awful lot of time because I misunderstood the LCD itself, rather than the microcontroller peripheral.

Memory-Mapped Peripherals

You have probably noticed the term "memory-mapped" peripherals in several places in the text. For those unfamiliar with the term, it describes the method by which the CPU communicates with the on-device peripheral modules. Memory-mapped peripherals have all control, input, and output registers located in address space. The advantage of this type of peripheral control is CPU simplicity. The CPU doesn't care a whit about the nature of the peripherals with which it communicates. It merely reads and writes values to a specific address. It also allows for functional scaling of devices without needing to modify the CPU. Memory-mapped peripherals are prevalent in von Neumann architectures.

The alternative to memory mapping is instruction mapping. In this method, each peripheral device has specific instructions associated with it. For example, a timer unit would have "start timer" and "clear timer" instructions as part of the device instruction set. This method, which is more prevalent in Harvard architectures, seems to be falling out of favor. There are still, however, some devices out there that use this approach.

The MSP430, being a memory-mapped device, has a very simple and straightforward architecture. Look through the disassembly of any file you have developed in the past. You will notice that the vast majority of instructions are reading or writing to the ALU, a CPU register, or some location in the address space. That is pretty much all the CPU does; move stuff around. The intelligence lies in the locations from and to which information is moved.

CHAPTER 7

Hardware Considerations

Embedded systems design is widely regarded as being a software development endeavor. Often, the hardware portion is overlooked. In my experience, hardware problems tend to consume more time than software problems in the debugging process. There are several reasons for this. First, most of our development time is spent writing the software, and we develop a certain level of comfort with it. This is especially true of designers coming from a software engineering background. This better understanding of the software tends to lead to more efficient debugging.

The second reason for the discrepancy in hardware/software debugging effort is that of tools. ROM simulation, high-end emulators, and software tools designed specifically for this purpose give the designer an awful lot of weapons in the battle against bugs. We can speed up and slow down execution rates, set breakpoints, and manually step through the code in search of any software misbehavior. In contrast, our hardware debugging tools usually consist of a good oscilloscope, a digital meter, and our wits. Also, the process of chasing down bugs is not usually as well-defined for hardware as it is for code.

Because of this, we need to be careful and thorough with the hardware portion of our designs. This chapter attempts to describe some of the more important issues and features related to hardware. It is by no means comprehensive, as I seem to discover new and unique hardware issues on every project I work on. However, I have included the basics, along with some '430-specific lessons I have learned.

The Datasheet

Any discussion of hardware must begin with the datasheet. The datasheet offers a good starting point for understanding the behavior of the device, and defines some design parameters. Some of my datasheet practices:

- Make sure you have the latest datasheet. The revision date is on the top of the first page. Go to *www.ti.com*, and download the most recent update.
- Keep a "working copy." Print out a copy of the device datasheet early in development, and scribble down notes on any problems and discoveries you make along the way. You will be surprised how many times you revisit the same issues on different designs, and those old notes will make short work of chasing down the new problems.
- Ignore page 1. This advice is not specific to TI parts, or even to microcontrollers in general. The first page of most datasheets is pure marketing stuff. It will describe some of the salient features, and give design parameters (current consumption, in the case of the '430) for specific, near-ideal conditions. Your design might or might not meet those parameters. The useful information you need might be in the datasheet, but it is not on the first page.
- Confirm critical values. If power consumption is among your primary concerns, build a prototype and measure the consumption. The datasheet will give some idea, but will not cover all combinations of peripherals, I/O, and supply voltages.
- Don't exceed datasheet limits. One of my favorite conversations to have with reps is about flash reprogramming cycles. I have had this same conversation with the application engineer or marketing representative from at least four different companies, when they come in to promote their micros:

Rep: "Our datasheet says 10,000 reprogramming cycles, but you can plan on 100,000."

Me: "So, you are willing to guarantee 100,000 at 3 or 4 sigma?"

Rep: "Uhhmm, no."

The lesson here is that, although you will typically see better performance than the min/max listed on the datasheet, don't count on it all the time. The manufacturer sets design limits where they are to account for the variation among devices, applications, and environments. Exceeding these limits in your design is asking for failures in the field.

Configuration

Power Supplies and Reset

Power supply design tends to be neglected by embedded developers. The approach of "Vcc is at 3 volts, it must be fine" is the source of many a bug in our field. Your biggest enemy here is noise. A noisy supply line can create unexpected resets and interrupts, disrupt register values, and foul up your crystal oscillators. Susceptibility to supply noise (and, in fact, noise to the device in general) is a particular weakness of the MSP430 family.

Along with the subject of power, grounding needs to be discussed. Many designs use a single, common ground for all signals, but this is not necessarily true for all designs. The MSP430 devices offer separate digital and analog grounds (labeled DVss and Avss on the datasheet). The analog supply and ground are only used in the analog side of the A/D converter, so you will only need to consider multiple grounds when measuring a floating analog signal. If at all possible, you are better off designing with a single ground, and tying DVss and AVss together externally. Unfortunately, this is not always possible. When using the grounds independently, be careful to keep them separate everywhere in the circuit, and be clear and careful about which ground each individual signal is referenced to. On the '430 devices, all of the signals except the A/D inputs are referenced to DVss. Since these A/D inputs are on the same pins as general purpose digital I/O on devices, your designs are restricted from externally multiplexing these signals, and using both functions for these pins, unless a common-ground scheme is used.

Along with supply lines, the '430 devices have a separate hardware reset pin, identified as ~RST/NMI on the datasheet. The hardware reset interrupt is described in chapter 3. This pin should be configured with an RC circuit or dedicated reset control device, so that the pin voltage is held low while the power supply stabilizes (see figure 7.x). The reset signal is digital, and should be referenced to DVss.

Crystals and External DCO

The connection required for external crystals is simple and straightforward. Use of loading capacitors varies by both device and crystal selection, so check the datasheets. Generally, however, the low-frequency crystals can be connected to most devices directly, without the need for decoupling capacitors. High frequency crystals tend to need the capacitors. If you are fortunate enough to have a true TTL or CMOS level oscillator, simply feed it to the XIN (or XT2IN) pin, turn the oscillator off with the OSCOff (or XT2Off) line, and float or ground XOUT (or XT2OUT).

The DCO can be run externally with the connection of a single resistor. I am not certain, however, why you would. It is possible to perform oscillator correction (for temperature and aging) with an external digital potentiometer using this configuration, but it is probably cheaper, easier, and more reliable to put an adequately accurate crystal on the device. I am certain that there is an application that will require this, but I have yet to see it. Suffice it to say that this is a single-element connection.

A/D Converters

A/D converter inputs need to be held relatively stable for brief periods of time. The external connections required for an A/D input are minimal. The converter has an internal sample and hold circuit, so you need not worry about that. The only external connection that I typically use is a decoupling capacitor, to keep high-frequency noise off the line. I honestly do not know how much performance gain it really buys, but it is another one of those inexpensive design habits that I have developed over time.

Performance Issues

Description of the general hardware performance of these devices can be summed up in a single word: **sensitive**. A trade-off for the very low power consumption, the '430 family of devices tend to react to any surge, noise, transient, or light breeze applied to the circuit. It is difficult to quantify this sensitivity, but I have experienced enough examples, both in my own designs and others, that I am comfortable in stating that these devices are less robust to errant signal conditions than most other controllers on the market.

I have a perfect example of this sensitivity at my bench as I write this. I am using a MSP430F149 to control several communications devices. The circuit is fed by a 3.3 VDC power supply, and an 8.000 MHz source. In order to trigger the code, an external interrupt is generated on port 1. Connected to the pin is about 16 inches of 22 gauge wire. My original intent was to touch this to supply to generate the interrupt. I quickly discovered, however, that the interrupt was being generated before I even touched the wire. I am able to generate the interrupt, consistently and repeatably, by waving my hand over the wire. The motion of my hand over the wire creates enough change in the surrounding magnetic field (which is notable, with all of the bench equipment running nearby) to create current in that wire, generating the interrupt. Seeing this phenomenon once or twice is not uncommon, but I find the rock-solid repeatability to be remarkable.

A second and particularly annoying offshoot of this sensitivity is that of brownout. If supply dips, but does not drop to ground, the device seems to enter a locked state. On a completed design, this should be well ironed out prior to manufacture, and usually is. I have found it particularly irritating in development, when my bench is covered in a nest of wires, cables, circuit boards, and scope probes. Bump the supply cable, and the whole thing locks up. The experienced engineer is now thinking "clean up your bench, and the problem will fix itself". While they are not incorrect, this tends to happen in pretty stable setups. The short version is, with this device, you will need to overdesign supply and reset circuits.

This sensitivity is not a problem if the designer expects it. It is difficult to completely design out these problems, but there are some pretty common steps you can take to improve the situation:

- Protect your power. If you are absolutely certain, beyond a shadow of a doubt, that there will be no noise on your supply lines, use several decoupling capacitors anyway. If you are unsure of your supply, design in some real filtering.
- Keep the reset line stable. Instability of the connection to the RST/NMI pin has created blood-pressure problems for many an embedded developer, the author included. There are some very good devices available, specifically designed to maintain microcontroller reset lines. Find one.
- Tie up the unused I/O pins. Floating inputs make wonderful antennae, allowing random signals onto the device. I prefer to tie these pins straight to ground, but that is certainly not the only method. There are various trade-offs between current consumption, power supply and ground stability, and parts count to be made on any design.
- Select your external interrupts carefully. The architecture allows the code to select individual pins for interrupts, and to enable and disable them on the fly. This can be a very useful feature for preventing unwanted interrupts.

Debugging Tools

Bench Equipment

A digital multimeter and a good oscilloscope are the fundamental "must haves" of hardware debugging. They are the tools that will give you the first glimpse into hardware behavior. My debugging setup actually has two meters. The first one has the common lead fixed to the system ground, so that I can check voltages with a single touch. The second meter is primarily used for checking resistances and continuity. I also have a couple of o-scopes at my bench, but the second one is something of a luxury. My primary scope is a high-end, 4-channel, 1-GHz digital scope, with a floppy drive and several different data I/O's (GPIB, serial). I have found that, when debug time rolls around, you can never have too powerful a scope.

Logic analyzers are tools that have really grown in functionality over the years. Older analyzers will provide bit, byte, and data bus streams, providing groups of 1's and 0's. In recent years, digital o-scope type functionalities, such as glitch detection, signal analysis, and deep trace memory have made their way into the logic analyzer market. Logic analyzers tend to be what I call "10 percent tools," because they are useful, or even critical, for a small percentage of the problems you are likely to encounter, such as race condition or bus contention. There are some very powerful analyzers available today, along with some low-cost, low-end PC-based tools. Because of the nature of the problems you will be hunting with this tool, the high-end models are almost always worth the extra cost.

There are a few other bits of equipment that are nice to have when developing and debugging. An obvious need is a solid DC power supply. Switching supplies are adequate for virtually any applications, as long as the switching noise isn't coming through the supply lines at any significant level. This is easy enough to check with the scope. A function generator, to use as a clock source, is another "nice to have," Like working with crystals, accuracy is among the most significant things to be aware of with the function generator.

Emulators

One of the best ways to view the internal interaction between device hardware and your code is through use of an emulator. Capabilities of emulation systems vary widely, depending on the manufacturer of the system and the target device. The advent of flash memory devices and JTAG (and BDMtype) debuggers has had a detrimental effect on the emulator market in recent years. TI offers a low-cost ($99, at time of printing) development tool, which is described in Appendix B. I have several of these at my bench. I use them for some formal development, but have found them especially useful for proof of concept work.

Emulators for the MSP430 family are developed and sold by Hitex (*www.hitex.com*). I have found it to be a reasonably good emulation system. It is feature-rich, and the accompanying software is as intuitive and useable as any other emulation software I have worked with. He biggest downside of the Hitex system is the cost. They offer several levels of system performance, typically ranging from $6,000 to $14,000. If you are shopping with your employers' checkbook (and have an understanding manager), this is not an unreasonable expense. For operations that expect to ship thousands (or millions) of whatever you are working on, the emulator will more than pay for itself, if used effectively. If you are a hobbyist, the emulator system is probably a bit too pricey.

The only real advice I have is to spend some time learning the features of the emulation system. It is a complex and powerful tool, but will require some time before you understand its features and functions well enough to get the most from them. Do not expect to be performing complicated and useful testing on the day you receive the emulation system.

CHAPTER 8

Addressing Modes

The MSP430 supports seven different addressing modes. For source operands, all seven are fully supported. For destination operands, four of the modes are directly supported, and two more can be emulated. The only "source only" mode is immediate mode, in which the source operand is a constant. This simply wouldn't make any sense for a destination. (Think about it—does the statement "7=9" make sense in any programming language?)

Some Notes on Instruction Notation:

Instructions in this chapter, and in Chapter 9, are described using the following notation:

```
<instruction>        src,dest        ;comments
```

where "src" is the source operand and "dest" is the destination operand. In single operand instructions, the source is omitted, and source and destination operands are both omitted in the few no operand instructions. Addressing modes are determined in the controller by the variables W(D) and W(S), respectively. W(D) is a single bit, indicating the presence or absence of a destination word, while W(S) is a bit pair, used to signify the source addressing mode. (The TI documentation indicates these variables by Ad and As, instead of W(D) and W(S). I have adopted the W notation, because I find it more descriptive.)

As in the rest of the book, notation in this chapter assumes register labels and constants defined in the file msp430x14x.h (or a similar file for other sub-families). These files are available from TI, either off their website or with development tools (including Kickstart). I strongly recommend inclusion and use of these files in any project. The standard names they define for registers and flags are used throughout the TI documentation, and this book.

Register Mode

Description: Register mode operations are the simplest and fastest operations performed by the processor. They are operations directly on the processor registers, R4 through R15, or on special function registers, such as the program counter or status register. For register source operands, W(S) = 00 and the source register is defined in the opcode. For register destination operands, W(D)=0 and the destination register is also defined in the opcode.

Advantages: Register mode is very efficient in terms of both instruction speed and code space. Register mode operations can be accomplished in a single clock cycle. It is also very straightforward, and easy to program in this mode.

Disadvantages: Register mode is limited to the twelve processor registers. If your application requires more than twelve distinct values (or fewer, if you need to use any 32-bit values), register mode alone will prove insufficient. Also, accessing anything in memory space (including RAM, hard-coded look up tables, or instruction memory) will require a different mode.

Uses: Register mode is a general-use tool. Any values which will be accessed or changed more than four or five times should be copied into the processor registers and accessed from there.

Examples:

Register contents before operation:

R15=0FF0 R14=5A5A R13=1000

Operation:

```
mov     R14,R15         ; Copies contents of R14 into R15
```

Register contents after operation:

R15=5A5A R14=5A5A R13=1000

Register contents before operation:

R15=0FF0 R14=5A5A R13=1000

Operation:

```
bis     R13,R15         ; Set bits in R15 based on R13
```

Register contents after operation:

R15=1FF0 R14=5A5A R13=1000

Register contents before operation:

R15=0FF0 R14=5A5A R13=1000

Operation:

```
clr.b   R14             ; Clear low byte of R14
```

Register contents after operation:

R15=5A5A R14=5A00 R13=1000

Register contents before operation:

R15=0FF0 R14=5A5A R13=1000

Operation:

```
mov     #0FFFFh,R15     ; Immediate/register mode combination
```

Register contents after operation:

R15=FFFF R14=5A5A R13=1000

Cycles Required:

Operands	2nd Operand Mode	Operator	Cycles	Length (words)
2	Register	Any	1*	1
2	Indexed, Symbolic or Absolute	Any	4	2
1	N/A	RRA, RRC, SWPB, or SXT	1	1
1	N/A	CALL or PUSH	5	1

*Register mode branch operations (mov Rn, PC) require 2 cycles.

Immediate Mode

Description: Immediate mode is used to assign constant values to registers or memory locations. Some specific values (0000h,0001h,0002h,0004h,0008h, and FFFFh) can be generated by the constant generators R2 and R3. In these cases, immediate mode behaves like register mode, and no data word is required. Otherwise, the constant is captured in a data word, which immediately follows the opcode.

Advantages and Disadvantages: Since immediate mode is primarily a utility function, it really doesn't have advantages and disadvantages, per se. It is more of a "use it when you need it" function.

Uses: Used to assign constant values in code. Most of the emulated instructions are immediate mode instructions.

Examples:

Register contents before operation:

R15=0FF0 R14=5A5A

Operation:

```
mov         #000F,R15            ; Move value into R15
```

Register contents after operation:

R15=000F R14=5A5A

Register contents before operation:

R15=0FF0 R14=5A5A

Operation:

```
clr         R15                  ; Clear R14
```

Register contents after operation:

R15=000F R14=0000

Operation:

```
bis         MC_2,&TACTL          ; Start timer A in continuous up mode
```

Operation:

```
call        FOO                  ; Call subroutine at label FOO
```

Cycles Required:

Operands	2nd Operand Mode	Operator	Cycles	Length (words)
2	Register	Any	2*	2
2	Indexed, Symbolic or Absolute	Any	5	3
1	N/A	RRA, RRC, SWPB, or SXT	N/A	N/A
1	N/A	CALL or PUSH	5	2

*If the destination is PC, this operation requires 3 cycles.

Indexed Mode

Description: Indexed mode commands are formatted as n(Rx), where n is a constant and Rx is one of the CPU registers. The absolute memory location n+(contents of Rx) is addressed. For indexed source operands, W(S) = 01 and the memory location is defined by the word immediately following the opcode. For indexed destination operands, W(D)=1 and the memory location is also defined by the word immediately following the opcode. In the case where both source and destination are indexed, the source word precedes the destination word.

Indexed mode creates opcodes identical to those created by symbolic and absolute modes. Indexed mode also serves as the destination emulation for indirect mode.

Advantages: Indexed mode is very useful for implementation of lookup tables. Define a constant memory location (name it something like Table_Start, for reference), and compute the table location that needs to be returned into a CPU register. The table value can then be returned with Table_Start(Rx).

Disadvantages: Indexed mode requires extra words in the instruction pipeline, in order to define the memory location to be addressed. It is also not the best approach for addressing memory at a fixed, known location.

Uses: Indexed mode is useful for applications such as lookups, when the location to be addressed is at a variable, computable distance form a known location in memory.

Examples:

Register contents before operation:

```
R15=011F R14=FFFF             Memory Location 0xF11F=5A5A
```

Operation:

```
mov      0F000h(R15),R14      ; Copies contents from memory to R14
```

Register contents after operation:

```
R15=5A5A      R14=5A5A        Memory Location 0xF11F=5A5A
```

Cycles Required:

Operands	2nd Operand Mode	Operator	Cycles	Length (words)
2	Register	Any	3	2
2	Indexed, Symbolic or Absolute	Any	6	3
1	N/A	RRA, RRC, SWPB, or SXT	4	2
1	N/A	CALL or PUSH	5	2

Symbolic Mode/Absolute Mode

Description: I have grouped symbolic and absolute modes together because they are essentially the same. These modes allow for the addressing of fixed memory locations, through assignment of labels to those locations. The difference between the two modes is in how the memory labels are used in the code. In symbolic mode, the label itself is used; and in absolute mode, it is preceded by a "&". (I suspect that TI put in the "ampersand mode" to appease C programmers, who use a similar structure for dereferencing of pointers.) Symbolic and absolute modes create opcodes identical to those created by indexed mode. The address is computed slightly differently by the assembler in each mode, but the end result is the same.

For symbolic and absolute source operands, W(S) = 01 and the memory location is defined by the word immediately following the opcode. For symbolic and absolute destination operands, W(D)=1 and the memory location is also defined by the word immediately following the opcode. In the case where both source and destination are indexed, the source word precedes the destination word.

Advantages: The assignment of labels to fixed memory locations is a practice that comes naturally to embedded programmers. The two modes allow for addressing either with or without the ampersand, so the developer can select the mode with which he or she is most comfortable.

Disadvantages: Symbolic and absolute modes require extra words in the instruction pipeline, in order to define the memory location to be addressed.

Uses: Any fixed location memory addressing, including use of RAM variables, should be performed with one of these modes.

Examples:

Memory contents before operation:

Location	FOO=1000	Location BAR=A5A5

Operation:

mov	FOO,BAR	; Copies contents of FOO into BAR (Symbolic)

Memory contents before operation:

Location	FOO=1000	Location BAR=1000

Memory contents before operation:

Location	FOO=1000	Location BAR=A5A5

Operation:

mov	&FOO,&BAR	; Copies contents of FOO into BAR (Absolute)

Memory contents before operation:

Location	FOO=1000	Location BAR=1000

Cycles Required:

Operands	2nd Operand Mode	Operator	Cycles	Length (words)
2	Register	Any	3	2
2	Indexed, Symbolic or Absolute	Any	6	3
1	N/A	RRA, RRC, SWPB, or SXT	4	2
1	N/A	CALL or PUSH	5	2

Indirect Mode/Indirect Autoincrement Mode

Description: These two modes are identical, except, as one would expect, the autoincrement feature increments the operand as part of the instruction. Incrementation is performed after memory access (i.e.postincrement). Indirect mode is similar to C language pointers. The format for operands is @Rn (Rn is a processor register), or @Rn+ for autoincrement mode. The data word addressed is located in the memory location pointed to by Rn. (E.g. if the word in R15 is 0xE5FF, the data addressed by @R15 is the data word in memory location E5FF.) For indirect source operands, W(S) = 10. For indirect autoincrement mode, W(S) = 11. Indirect mode is not valid for destination operands, but can be emulated with the indexed mode format 0(Rn). Indirect autoincrement mode is also emulated with indexed mode, but requires an extra statement (inc Rn).

Advantages: The ability to point to a memory location and increment the pointer in a single instruction is a very useful instruction, if used properly.

Disadvantages: Like C pointers, indirect mode can be a source of confusion. It is easy for the beginning developer to mentally mix up values and pointers to values in their mind. Be wary. Also, the indirect modes require extra words in the instruction pipeline, in order to define the memory location to be addressed.

Uses: C compilers will use this mode extensively, especially when implementing pointers. It is very handy for iterative memory access, such as array operations.

Examples:

Register contents before operation:

R15=0FF0 R14=EA5A   Memory location 0xEA5A = 1234

Operation:

```
mov     @R14,R15        ; Move value into R15
```

Register contents after operation:

R15=1234 R14=EA5A   Memory location 0xEA5A = 1234

Register contents before operation:

R15=0FF0 R14=EA5A Memory location 0xEA5A = 1234

Operation:

```
mov      @R14+,R15        ; Move value into R15
```

Register contents after operation:

R15=1234 R14=EA5B Memory location 0xEA5A = 1234

Register contents before operation:

R14=02FF Memory location 0x02FF = 1234

Operation:

```
clr      @R14             ; Clear RAM location
```

Register contents after operation:

R14=02FF Memory location 0x02FF = 0000

Register contents before operation:

R14=02FF Memory location 0x02FF = 1234

Operation:

```
clr      @R14+            ; Clear RAM location
```

Register contents after operation:

R14=0300 Memory location 0x02FF = 0000

Operation:

```
pop      R12        ; pop is an emulated autoincrement instruction
```

Cycles Required:

Operands	2nd Operand Mode	Operator	Cycles	Length (words)
2	Register	Any	2	1
2	Indexed, Symbolic or Absolute	Any	5	2
1	N/A	RRA, RRC, SWPB, or SXT	3	1
1	N/A	CALL or PUSH	4	1

*Indirect to register mode branch operations (mov @Rn, PC) require 3 cycles.

Instruction Set Orthogonality

The ability to use different addressing modes for both source and destination is referred to as instruction *orthogonality*. The '430 is considered to be fully orthogonal, since any instruction can effectively use any addressing mode for both source and destination operands. This flexibility is something of a rarity in small microcontrollers.

Full orthogonality has several advantages. It allows the programmer to write very compact code. Many operations which require multiple statements in other controllers can be accomplished with one properly addressed command in the '430. Also, and perhaps most importantly, this is a very compiler-friendly feature. C constructs, such as pointers, implement much more readily in the '430 than in most other microcontrollers. The downside, however, is added complexity. In order to implement a fully orthogonal set, linear instruction timing is sacrificed (i.e., instructions can take anywhere from 1 to 6 clock cycles, depending on addressing).

The complexity can really be an advantage or a hindrance, depending on the developer. The '430 certainly requires the coder to hold a deeper understanding of the device than most competing platforms. The combination of orthogonality, code mapped in the same space as RAM, and multiple on-chip peripherals requires greater mental discipline than simply shuffling data between RAM and a single working register. In short, this device gives the developer plenty of rope.

CHAPTER 9

Instruction Set

This chapter describes the MSP430 instruction set. Quick reference tables categorize the instructions by operation type. The instructions are then individually described, with their effect on flags, and some common examples.

The MSP430 instruction set consists of 27 base opcodes, some of which can be called as either byte or word instructions. Additionally, TI has documented 24 emulated instructions, which are supported by assemblers and compilers, but have no dedicated opcode associated with their mnemonic. Byte and word instructions are called via a suffix of .B or .W to the mnemonic. For example, MOV.B is a byte move, and MOV.W is a 16-bit word move. If the suffix is omitted, the mnemonic is interpreted to be a word instruction. Bit and flow control operations do not support a suffix.

The use of emulated instructions might be viewed as optional or extraneous to the basic design and use of the device. They are, in fact, quite necessary in many cases. There are several cases where base instructions have complementary emulated instructions. The best example of this is PUSH (base) and POP (emulated). The right handed roll operations are base, while the left handed are emulated. C compilers (and most assembly programmers) will rely heavily on these emulated instructions.

Arithmetic Instructions

Mnemonic	Description	Emulation
ADC(.B or .W) dest	Add carry to destination	ADDC(.B or.W) #0,dest
ADD(.B or .W) src,dest	Add source to destination	Not an emulated instruction
ADDC(.B or .W) src,dest	Add source and carry to destination	Not an emulated instruction
DADC (.B or .W) dest	Decimal add carry to destination	DADD(.B or .W), #0,dest
DADD (.B or .W) src,dest	Decimal add source and carry to destination	Not an emulated instruction
DEC (.B or .W) dest	Decrement destination	SUB(.B or .W) #1,dest
DECD (.B or .W) dest	Decrement destination twice	SUB(.B or .W) #2,dest
INC(.B or .W) dest	Increment destination	ADD(.B or .W) #1,dest
INCD(.B or .W) dest	Increment destination twice	ADD(.B or .W) #2,dest
SBC(.B or .W) dest	Subtract carry from destination	SUBC(.B or .W) #0,dest
SUB(.B or .W) src,dest	Subtract source from destination	Not an emulated instruction
SUBC(.B or .W) src,dest	Subtract source and borrow* from destination	Not an emulated instruction

*borrow is defined as NOT carry

Logical and Register Control Instructions

Mnemonic	Description	Emulation
AND(.B or .W) src,dest	AND source with destination	Not an emulated instruction
BIC(.B or .W) src,dest	Clear bits in destination	Not an emulated instruction
BIS(.B or .W) src,dest	Set bits in destination	Not an emulated instruction
BIT(.B or .W) src,dest	Test bits in destination	Not an emulated instruction
INV(.B or .W) dest	Invert bits in destination	XOR(.B or .W) #0FFFFh,dest
RLA(.B or .W) dest	Roll destination left	ADD(.B or .W) dest,dest
RLC(.B or .W) dest	Roll destination left through carry	ADDC(.B or .W) dest,dest
RRA(.B or .W) dest	Roll destination right	Not an emulated instruction
RRC(.B or .W) dest	Roll destination right through (from) carry	Not an emulated instruction
SWPB dest (word only)	Swap bytes in destination	Not an emulated instruction
SXT dest (word only)	Sign extend destination	Not an emulated instruction
XOR(.B or .W) dest	XOR source with destination	Not an emulated instruction

Data Instructions

Mnemonic	Description	Emulation
CLR(.B or .W) dest	Clear destination	MOV(.B or .W) #0,dest
CLRC	Clear carry flag	BIC #1,SR
CLRN	Clear negative flag	BIC #4,SR
CLRZ	Clear zero flag	BIC #2,SR
CMP(.B or .W) src,dest	Compare source to destination	Not an emulated instruction
MOV(.B or .W) src,dest	Move source to destination	Not an emulated instruction
POP(.B or .W) dest	Pop item from stack to dest	MOV(.B or .W) @SP+,dest
PUSH(.B or .W) dest	Push dest to stack	Not an emulated instruction
SETC	Set carry flag	BIS #1,SR
SETN	Set negative flag	BIS #4,SR
SETZ	Set zero flag	BIS #2,SR
TST(.B or .W) dest	Test destination	CMP(.B or .W) #0,dest

Program Flow Control Instructions

Mnemonic	Description	Emulation
BR dest	Branch to destination	MOV dest,PC
CALL dest	Subroutine call to destination	Not an emulated instruction
DINT	Disable interrupts	BIC #8,SR
EINT	Enable interrupts	BIS #8,SR
JC (or JHS) label	Jump to label if carry flag is set	Not an emulated instruction
JGE label	Jump to label if greater than or equal *	Not an emulated instruction
JL label	Jump to label if less than *	Not an emulated instruction
JMP label	Jump to label unconditionally	Not an emulated instruction
JN label	Jump to label if negative flag is set	Not an emulated instruction
JNC (or JLO) label	Jump to label if carry flag is reset	Not an emulated instruction
JNZ (or JNE) label	Jump to label if zero flag is reset	Not an emulated instruction
JZ (or JEQ) label	Jump to label if zero flag is set	Not an emulated instruction
NOP	No operation	MOV #0,#0
RET	Return from subroutine	MOV @SP+,PC
RETI	Return from interrupt	Not an emulated instruction

*jump is made based on flag values from a previous CMP statement

Core Instructions

ADD (.B or .W) src,dest   Add source to destination

Description: The source operand is added to the destination oper and, and the result is placed in the destination. The value in source is preserved.

Operation: dest = dest + source

Opcode Structure:

0	1	0	1	R(S)	R(S)	R(S)	R(S)	W(D)	B/W	W(S)	W(S)	R(D)	R(D)	R(D)	R(D)

R(S): Source Register (0 if not register operation)
W(D): 1=Destination word (index, symbolic, or absolute)
: 0=No destination word (register mode)
B/W: 1=Byte Instruction
0=Word Instruction
W(S): 00=No source word (register mode)
01=Index, symbolic, or absolute mode for source
10=Indirect register mode
11=Indirect autoincrement or immediate mode
R(D): Destination Register (0 if not register operation)

Status Flags: All flags affected normally

Examples:

```
ADD     #3Ah,R15     ; Add the value 3A to R15
ADD     R15,&MAC     ; Add the value in R15  to multiplier SFR
ADD     @R4,R7       ; Add the contents of the location
                       pointed to by R4 to R7
JC      FOO          ; If carry flag is set by operation, jump
                       to FOO
```

ADDC (.B or .W) src,dest   Add source and carry to destination

Description: The source operand and carry flag are added to the destination operand, and the result is placed in the destination. The value in source is preserved.

Operation: dest = dest + source + C

Opcode Structure:

0	1	1	0	R(S)	R(S)	R(S)	R(S)	W(D)	B/W	W(S)	W(S)	R(D)	R(D)	R(D)	R(D)

R(S): Source Register (0 if not register operation)
W(D): 1=Destination word (index, symbolic, or absolute)
: 0=No destination word (register mode)
B/W: 1=Byte Instruction
0=Word Instruction
W(S): 00=No source word (register mode)
01=Index, symbolic, or absolute mode for source
10=Indirect register mode
11=Indirect autoincrement or immediate mode
R(D): Destination Register (0 if not register operation)

Status Flags: All flags affected normally

Examples:

```
ADDC    #3Ah,R15      ; Add the value 3A to R15
ADDC    R15,&MAC      ; Add the value in R15  to multiplier SFR
ADDC    @R4,R7        ; Add the contents of the location
                        pointed to by R4 to R7
JC      FOO           ; If carry flag is set by operation, jump
                        to FOO
```

AND (.B or .W) src,dest — Logical AND bits in source and destination

Description: The bits in the source operand are logically ANDed with the bits in the destination operand, and the result is placed in the destination. The value in source is preserved.

Operation: dest = dest AND src

Opcode Structure:

1	1	1	1	R(S)	R(S)	R(S)	R(S)	W(D)	B/W	W(S)	W(S)	R(D)	R(D)	R(D)	R(D)

R(S): Source Register (0 if not register operation)
W(D): 1=Destination word (index, symbolic, or absolute)
: 0=No destination word (register mode)
B/W: 1=Byte Instruction
0=Word Instruction
W(S): 00=No source word (register mode)
01=Index, symbolic, or absolute mode for source
10=Indirect register mode
11=Indirect autoincrement or immediate mode
R(D): Destination Register (0 if not register operation)

Status Flags:
Z: Set if result=0, reset otherwise
C: NOT Z
N: Takes value of result MSB
V: Reset

Examples:

```
AND     @R4,R7       ; AND the contents of the location
                       pointed to by R4 to R7
AND     #00FFh,R12   ; Mask off the top byte of R12
```

BIC (.B or .W) src,dest Clear selected bits in destination

Description: Specific bits in the destination, as determined by the set bits in the source, are cleared. The value in source is preserved.

Operation: dest = dest AND NOT(src)

Opcode Structure:

1	1	0	0	R(S)	R(S)	R(S)	R(S)	W(D)	B/W	W(S)	W(S)	R(D)	R(D)	R(D)	R(D)

R(S): Source Register (0 if not register operation)
W(D): 1=Destination word (index, symbolic, or absolute)
: 0=No destination word (register mode)
B/W: 1=Byte Instruction
0=Word Instruction
W(S): 00=No source word (register mode)
01=Index, symbolic, or absolute mode for source
10=Indirect register mode
11=Indirect autoincrement or immediate mode
R(D): Destination Register (0 if not register operation)

Status Flags: Status flags are unaffected by this operation.

Examples:

```
BIC   #0030h,&TACTL     ; Stop timer A
BIC   #00FFh,R12        ; Mask off the top byte of R12
```

BIS (.B or .W) src,dest   Set selected bits in destination

Description: Specific bits in the destination, as determined by the set bits in the source, are set. The value in source is preserved.

Operation: dest = dest OR src

Opcode Structure:

1	1	0	1	R(S)	R(S)	R(S)	R(S)	W(D)	B/W	W(S)	W(S)	R(D)	R(D)	R(D)	R(D)

R(S): Source Register (0 if not register operation)
W(D): 1=Destination word (index, symbolic, or absolute)
: 0=No destination word (register mode)
B/W: 1=Byte Instruction
0=Word Instruction
W(S): 00=No source word (register mode)
01=Index, symbolic, or absolute mode for source
10=Indirect register mode
11=Indirect autoincrement or immediate mode
R(D): Destination Register (0 if not register operation)

Status Flags: Status flags are unaffected by this operation.

Examples:

```
BIS   #0030h,&TACTL     ; Start timer A in up/down mode
BIS   #00FFh,R7         ; Set the low byte of R7
```

BIT (.B or .W) src,dest — Test selected bits in destination

Description: Specific bits in the destination, as determined by the set bits in the source, are tested. The values in source and destination are both preserved. This operation and the AND operation affect flags identically.

Operation: dest AND src

Opcode Structure:

1	0	1	1	R(S)	R(S)	R(S)	R(S)	W(D)	B/W	W(S)	W(S)	R(D)	R(D)	R(D)	R(D)

R(S): Source Register (0 if not register operation)
W(D): 1=Destination word (index, symbolic, or absolute)
: 0=No destination word (register mode)
B/W: 1=Byte Instruction
0=Word Instruction
W(S): 00=No source word (register mode)
01=Index, symbolic, or absolute mode for source
10=Indirect register mode
11=Indirect autoincrement or immediate mode
R(D): Destination Register (0 if not register operation)

Status Flags: Z: Set if result=0, reset otherwise
C: NOT Z
N: Takes value of result MSB
V: Reset

Example:

```
BIT  #0040h,R12     ; is bit 6 of R12 set?
JZ   FOO            ; If yes, jump to label FOO
```

CALL dest — Subroutine call

Description: A subroutine call is made to destination, and the current program counter is pushed to the stack. Destination can be anywhere in the 64k memory space.

Operation: **PUSH** PC
PC = dest

Opcode Structure:

0	0	0	1	0	0	1	0	1	B/W	W(S)	W(S)	R(D)	R(D)	R(D)	R(D)

B/W: 0=Word Instruction (no byte mode for this operation)
W(S): 00=No source word (register mode)
01=Index, symbolic, or absolute mode for source
10=Indirect register mode
11=Indirect autoincrement or immediate mode
R(D): Destination Register (0 if not register operation)

Status Flags: Status flags are unaffected by this operation.

Example:

```
CALL    @R7     ; Call to address in word pointed to by R7
CALL    #FOO    ; Call to label FOO
```

CMP (.B or .W) src,dest Compare source to destination

Description: The source is subtracted from the destination. The values in source and destination are both preserved. This operation and the SUB operation affect flags identically.

Operation: dest-src

Opcode Structure:

1	0	0	1	R(S)	R(S)	R(S)	R(S)	W(D)	B/W	W(S)	W(S)	R(D)	R(D)	R(D)	R(D)

R(S): Source Register (0 if not register operation)
W(D): 1=Destination word (index, symbolic, or absolute)
: 0=No destination word (register mode)
B/W: 1=Byte Instruction
0=Word Instruction
W(S): 00=No source word (register mode)
01=Index, symbolic, or absolute mode for source
10=Indirect register mode
11=Indirect autoincrement or immediate mode
R(D): Destination Register (0 if not register operation)

Status Flags:
Z: Set if result=0, reset otherwise
C: Set if dest+ NOT.src +1 produces a carry, reset otherwise
N: Set if src >= dest, reset otherwise
V: Set on arithmetic overflow

Example:

```
CMP     FOO,R12     ; is R12=FOO?
JZ      BAR         ; If yes, jump to label BAR
```

DADD (.B or .W) src,dest — Decimal (BCD) add source and carry to destination

Description: The source and carry flag are added, in BCD format, to the destination.

Operation: dest(BCD) = dest (BCD) + src (BCD) + C

Opcode Structure:

1	0	1	0	R(S)	R(S)	R(S)	R(S)	W(D)	B/W	W(S)	W(S)	R(D)	R(D)	R(D)	R(D)

R(S): Source Register (0 if not register operation)
W(D): 1=Destination word (index, symbolic, or absolute)
: 0=No destination word (register mode)
B/W: 1=Byte Instruction
0=Word Instruction
W(S): 00=No source word (register mode)
01=Index, symbolic, or absolute mode for source
10=Indirect register mode
11=Indirect autoincrement or immediate mode
R(D): Destination Register (0 if not register operation)

Status Flags:
Z: Set if result=0, reset otherwise
C: Set if result>9999 (word) or result>99 (byte)
N: Takes value of MSB after operation
V: Undefined

Example:

```
DADD.B   R7,R8    ; add two decimals in R7 to R8
JC       BAR      ; If result >99, jump to BAR
```

JC dest — Jump if carry

Description: A jump to the destination is made if the carry flag is set. The jump can be made up to 511 words backwards or 512 words forwards in the code.

Operation: if C=1, PC = PC+2*offset
Else, perform a NOP

Opcode Structure:

0	0	1	0	1	1	Sign	Ofst	Ofst	Ofst	Ofst	Ofst	Ofst	Ofst	Ofst	Ofst

Sign = Sign bit for offset
Ofst = Address offset for jump

Status Flags: Status flags are unaffected by this operation.

Example:

```
ADDC      @R4,R7     ; Add the contents of the location
                       pointed to by R4 to R7
JC        FOO        ; If carry flag is set by operation, jump
                       to FOO

DADD.B    R7,R8      ; add two decimals in R7 to R8
JC        BAR        ; If result >99, jump to BAR
```

JGE dest — Jump if greater than or equal to

Description: A jump to the destination is made if the overflow flag and negative flag have the same value. The jump can be made up to 511 words backwards or 512 words forwards in the code.

Operation: if N.XOR.V=0, PC = PC+2*offset
Else, perform a NOP

Opcode Structure:

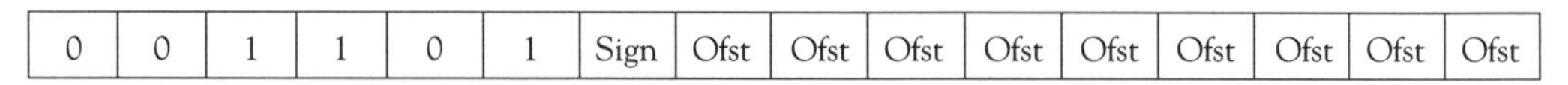

0	0	1	1	0	1	Sign	Ofst	Ofst	Ofst	Ofst	Ofst	Ofst	Ofst	Ofst	Ofst

Sign = Sign bit for offset
Ofst = Address offset for jump

Status Flags: Status flags are unaffected by this operation.

Examples:

```
CMP     R4,R7          ; Compare R4 to R7
JGE     FOO            ; If R7>=R6,  jump to FOO

CMP     R5,#0FF00h     ; Compare R5 to constant
JGE     BAR            ; If R5<constant,  jump to BAR
```

JL dest — Jump if less than

Description: A jump to the destination is made if the overflow flag and negative flag have complementary values. The jump can be made up to 511 words backwards or 512 words forwards in the code.

Operation: if N.XOR.V−1, PC − PC+2*offset
Else, perform a NOP

Opcode Structure:

0	0	1	1	1	0	Sign	Ofst	Ofst	Ofst	Ofst	Ofst	Ofst	Ofst	Ofst	Ofst

Sign = Sign bit for offset
Ofst = Address offset for jump

Status Flags: Status flags are unaffected by this operation.

Examples:

```
CMP     R4,R7         ; Compare R4 to R7
JL      FOO           ; If R7<R6,  jump to FOO

CMP     R5,#0FF00h    ; Compare R5 to constant
JL      BAR           ; If R5>=constant,  jump to BAR
```

JMP dest — Unconditional jump

Description: A jump to the destination is made. The jump can be made up to 511 words backwards or 512 words forwards in the code.

Operation: PC = PC+2*offset

Opcode Structure:

0	0	1	0	0	0	Sign	Ofst	Ofst	Ofst	Ofst	Ofst	Ofst	Ofst	Ofst	Ofst

Sign = Sign bit for offset
Ofst = Address offset for jump

Status Flags: Status flags are unaffected by this operation.

Examples:

```
JMP     FOO     ; Unconditional jump to FOO
```

JN dest — Jump if negative

Description: A jump to the destination is made if the negative flag is set. The jump can be made up to 511 words backwards or 512 words forwards in the code.

Operation: if N=1, PC = PC+2*offset
Else, perform a NOP

Opcode Structure:

0	0	1	1	0	0	Sign	Ofst	Ofst	Ofst	Ofst	Ofst	Ofst	Ofst	Ofst	Ofst

Sign = Sign bit for offset
Ofst = Address offset for jump

Status Flags: Status flags are unaffected by this operation.

Examples:

```
SUB     R4,R7     ; Subtract R7 from R4
JN      FOO       ; If result is negative,  jump to FOO
```

JNC dest — Jump if carry not set

Description: A jump to the destination is made if the carry flag is reset. The jump can be made up to 511 words backwards or 512 words forwards in the code.

Operation: if C=0, PC = PC+2*offset
Else, perform a NOP

Opcode Structure:

0	0	1	0	1	0	Sign	Ofst	Ofst	Ofst	Ofst	Ofst	Ofst	Ofst	Ofst	Ofst

Sign=Sign bit for offset
Ofst = Address offset for jump

Status Flags: Status flags are unaffected by this operation.

Example:

```
ADDC     @R4,R7     ; Add the contents of the location
                      pointed to by R4 to R7
JNC      FOO        ; If carry flag is not set by operation,
                      jump to FOO

DADD.B   R7,R8      ; add two decimals in R7 to R8
JNC      BAR        ; If result <99, jump to BAR
```

JNE or **JNZ** dest — Jump if not equal or jump if not zero

Description: A jump to the destination is made if the zero flag is reset. The jump can be made up to 511 words backwards or 512 words forwards in the code.

Operation: if Z=0, PC = PC+2*offset
Else, perform a NOP

Opcode Structure:

0	0	1	0	0	0	Sign	Ofst	Ofst	Ofst	Ofst	Ofst	Ofst	Ofst	Ofst	Ofst

Sign = Sign bit for offset
Ofst = Address offset for jump

Status Flags: Status flags are unaffected by this operation.

Example:

```
    CMP     R10,R11         ; Compare R10 to R11
    JNE     BAR             ; If R10<>R11,jump to BAR

    CMP     R8,#0FF00h      ; Compare R8 to constant
    JNE     FOO             ; If R5<>constant,  jump to FOO
```

JZ or **JEQ** dest — Jump if Zero or Jump if equal

Description: A jump to the destination is made if the zero flag is set. The jump can be made up to 511 words backwards or 512 words forwards in the code.

Operation: if Z=1, PC = PC+2*offset
Else, perform a NOP

Opcode Structure:

0	0	1	0	0	1	Sign	Ofst	Ofst	Ofst	Ofst	Ofst	Ofst	Ofst	Ofst	Ofst

Sign= Sign bit for offset
Ofst = Address offset for jump

Status Flags: Status flags are unaffected by this operation.

Example:

```
        MOV    #5,R7           ; initialize counter
FOO     ADD    @R6+,R5         ; Accumulate next sample in R5
        DEC    R7              ; Check for last sample
        JZ     BAR             ; If final sample, jump to label BAR
        JMP    FOO             ; else, loop to FOO
```

MOV (.B or .W) src,dest Move source to destination

Description: The source is copied to the destination. The value in source is preserved.

Operation: dest=src

Opcode Structure.

1	0	0	0	R(S)	R(S)	R(S)	R(S)	W(D)	B/W	W(S)	W(S)	R(D)	R(D)	R(D)	R(D)

R(S): Source Register (0 if not register operation)
W(D): 1=Destination word (index, symbolic, or absolute)
: 0=No destination word (register mode)
B/W: 1=Byte Instruction
0=Word Instruction
W(S): 00=No source word (register mode)
01=Index, symbolic, or absolute mode for source
10=Indirect register mode
11=Indirect autoincrement or immediate mode
R(D): Destination Register (0 if not register operation)

Status Flags: Status flags are unaffected by this operation.

Example:

```
MOV     #00FAh,R15   ; load constant into R15

MOV     @R14+,R4     ; Move contents of address in R14 to
                       R4, inc R14
```

PUSH (.B or .W) src   Push source to top of stack

Description: The stack pointer is decremented by two, and the source word is copied to the new TOS (top of stack) location.

Operation: SP=SP-2
@SP = src

Opcode Structure:

0	0	0	1	0	0	1	0	0	B/W	W(S)	W(S)	R(D)	R(D)	R(D)	R(D)

B/W: 1=Byte Instruction
0=Word Instruction

W(S): 00=No source word (register mode)
01=Index, symbolic, or absolute mode for source
10=Indirect register mode
11=Indirect autoincrement or immediate mode

R(D): Destination Register (0 if not register operation)

Status Flags: Status flags are unaffected by this operation.

Example:

```
PUSH    SR      ; Status register is pushed to stack
PUSH    R12     ; Push R12 to stack
```

RETI Return from interrupt

Description: Returns program flow from ISR to previous address. See Chapter 3 for description.

Operation:
POP SR
POP PC

Opcode Structure:

0	0	0	1	0	0	1	1	0	0	0	0	0	0	0	0

Status Flags: Status flags are restored from stack

Example:

```
RETI          ; return from interrupt
```

RRA (.B or .W) dest Roll right arithmetically

Description: All bits in the destination are shifted right one bit location. The MSB is preserved, and the LSB is shifted into the carry flag.

Operation: dest (MSB) preserved
dest (MSB)=>dest(MSB-1)
:
:
dest (LSB+1)=>dest(LSB)
dest (LSB)=>C

Opcode Structure:

0	0	0	1	0	0	0	1	0	B/W	W(S)	W(S)	R(D)	R(D)	R(D)	R(D)

B/W: 1=Byte Instruction
0=Word Instruction

W(S): 00=No source word (register mode)
01=Index, symbolic, or absolute mode for source
10=Indirect register mode
11=Indirect autoincrement or immediate mode

R(D): Destination Register (0 if not register operation)

Status Flags: Z: Set if result=0, reset otherwise
C: Loaded from LSB
N: Set if result<0, reset otherwise
V: Reset

Example:

```
      MOV   #4,R7       ; initialize counter for divide by 32
FOO   RRA   R5          ; Divide R5 by 2
      DEC   R7          ; Check for last divide
      JZ    BAR         ; If final sample, jump to label BAR
      JMP   FOO         ; else, loop to FOO
```

RRC (.B or .W) dest Roll right through carry

Description: All bits in the destination are shifted right one bit location. The carry flag is shifted into the MSB, and the LSB is shifted into the carry flag.

Operation:
C=>dest (MSB)
dest (MSB)=>dest(MSB-1)
:
:
dest (LSB+1)=>dest(LSB)
dest (LSB)=>C

Opcode Structure:

0	0	0	1	0	0	0	0	0	B/W	W(S)	W(S)	R(D)	R(D)	R(D)	R(D)

B/W: 1=Byte Instruction
0=Word Instruction
W(S): 00=No source word (register mode)
01=Index, symbolic, or absolute mode for source
10=Indirect register mode
11=Indirect autoincrement or immediate mode
R(D): Destination Register (0 if not register operation)

Status Flags:
Z: Set if result=0, reset otherwise
C: Loaded from LSB
N: Set if result<0, reset otherwise
V: Set if dest>0 and C=1, reset otherwise

Example:

```
FOO  RRC  R6        ; This set of commands shifts
     RRC  R5        ; 64 bits of data through the
     RRC  R4        ; processor registers and into the
     RRC  &P1OUT    ; Port 1 output register
```

SUB (.B or .W) src,dest — Subtract source from destination

Description: The source is subtracted from the destination. The value in source is preserved. This operation and the CMP operation affect flags identically.

Operation: dest = dest-src

Opcode Structure:

1	0	0	0	R(S)	R(S)	R(S)	R(S)	W(D)	B/W	W(S)	W(S)	R(D)	R(D)	R(D)	R(D)

R(S): Source Register (0 if not register operation)
W(D): 1=Destination word (index, symbolic, or absolute)
: 0=No destination word (register mode)
B/W: 1=Byte Instruction
0=Word Instruction
W(S): 00=No source word (register mode)
01=Index, symbolic, or absolute mode for source
10=Indirect register mode
11=Indirect autoincrement or immediate mode
R(D): Destination Register (0 if not register operation)

Status Flags:
Z: Set if result=0, reset otherwise
C: Set if dest+ NOT.src +1 produces a carry, reset otherwise
N: Set if src >= dest, reset otherwise
V: Set on arithmetic overflow

Example:

```
SUB   #4,R12       ; Subtract 4 from R12
JN    BAR          ; If R12 was <4 prior to op, jump to BAR
```

SUBC (.B or .W) src,dest   Subtract source and borrow from destination

Description: The source and borrow are subtracted from the destination. The value in source is preserved. Borrow is defined as the complement of the carry flag. This operation and the CMP operation affect flags identically.

Operation: dest = dest-src – 1 + C

Opcode Structure:

0	1	1	1	R(S)	R(S)	R(S)	R(S)	W(D)	B/W	W(S)	W(S)	R(D)	R(D)	R(D)	R(D)

R(S): Source Register (0 if not register operation)
W(D): 1=Destination word (index, symbolic, or absolute)
: 0=No destination word (register mode)
B/W: 1=Byte Instruction
0=Word Instruction
W(S): 00=No source word (register mode)
01=Index, symbolic, or absolute mode for source
10=Indirect register mode
11=Indirect autoincrement or immediate mode
R(D): Destination Register (0 if not register operation)

Status Flags: Z: Set if result=0, reset otherwise
C: Set if dest+ NOT.src +1 produces a carry, reset otherwise
N: Set if src >= dest, reset otherwise
V: Set on arithmetic overflow

Example:

```
SUBC    #4,R12    ; Subtract 4 from R12
JN      BAR       ; If R12 was <4 prior to op, jump to BAR
```

SWPB dest — Swap Bytes

Description: The high byte and the low byte in the destination word are swapped.

Operation:

temp=dest(high byte)
dest(high byte) = dest(low byte)
dest(low byte) = temp

Opcode Structure:

0	0	0	1	0	0	0	0	1	B/W	W(S)	W(S)	R(D)	R(D)	R(D)	R(D)

B/W: 0=Word Instruction (no byte mode for this operation)
W(S): 00=No source word (register mode)
01=Index, symbolic, or absolute mode for source
10=Indirect register mode
11=Indirect autoincrement or immediate mode
R(D): Destination Register (0 if not register operation)

Status Flags: Status flags are unaffected by this operation.

Example:

```
SWPB    R12           ; Divide by 256 by swapping bytes
AND     #00FFh,R12    ; and masking off top byte
```

SXT dest — Sign extend destination

Description: The sign of the low byte is copied into the high byte

Operation: dest(bits 8-15) = dest (bit 7)

Opcode Structure:

0	0	0	1	0	0	0	0	1	B/W	W(S)	W(S)	R(D)	R(D)	R(D)	R(D)

B/W: 0=Word Instruction (no byte mode for this operation)
W(S): 00=No source word (register mode)
01=Index, symbolic, or absolute mode for source
10=Indirect register mode
11=Indirect autoincrement or immediate mode
R(D): Destination Register (0 if not register operation)

Status Flags:
Z: Set if result=0, reset otherwise
C: NOT Z
N: Set if result<0, reset otherwise
V: Reset

Example:

```
MOV.B    &P2IN,R10;   ; Load R10 with 8-bit value from Port 2
SXT      R10          ; Sign extend for processing
```

XOR (.B or .W) src,dest — Logical XOR bits in source and destination

Description: The bits in the source operand are logically XORed with the bits in the destination operand, and the result is placed in the destination. The value in source is preserved.

Operation: dest = dest XOR src

Opcode Structure:

1	1	1	0	R(S)	R(S)	R(S)	R(S)	W(D)	B/W	W(S)	W(S)	R(D)	R(D)	R(D)	R(D)

R(S): Source Register (0 if not register operation)
W(D): 1=Destination word (index, symbolic, or absolute)
: 0=No destination word (register mode)
B/W: 1=Byte Instruction
0=Word Instruction
W(S): 00=No source word (register mode)
01=Index, symbolic, or absolute mode for source
10=Indirect register mode
11=Indirect autoincrement or immediate mode
R(D): Destination Register (0 if not register operation)

Status Flags:
Z: Set if result=0, reset otherwise
C: NOT Z
N: Takes value of result MSB
V: Set if both operands are negative

Examples:

```
XOR    @R4,R7       ; XOR the contents of the location
                      pointed to by R4 to R7
XOR    #00FFh,R12   ;Toggle the bits in the low byte of R12
```

Emulated Instructions

BR dest — Branch to destination

Description: Unconditional branch operation to any location in memory.
Operation: PC = dest
Emulation MOV dest,PC
Examples:

```
BR    #FOO    ; Branch to label FOO
BR    R12     ; Branch to address contained in R12
BR    @R12    ; Branch to address contained in word pointed
              to by R12
```

CLR (.B or .W) dest — Clear destination

Description: Destination is set to zero.
Operation: dest = 0
Emulation: MOV(.B or .W) #0,dest
Example:

```
CLR   R11     ; Clears R12
```

CLRC — Clear carry flag

Description: Carry flag is reset.
Operation: C = 0
Emulation BIC #1,SR
Example:

```
CLRC          ; Clears carry flag
```

CLRN — Clear negative flag

Description: Negative flag is reset.
Operation: N = 0
Emulation BIC #4,SR
Example:

```
CLRN          ; Clears negative flag
```

CLRZ Clear zero flag

Description: Zero flag is reset.

Operation: Z = 0

Emulation BIC #2,SR

Example:

```
CLRZ                    ; Clears zero flag
```

DADC (.B or .W) dest Add carry decimally to destination

Description: The carry flag is added to the destination in decimal (BCD) format.

Operation: dest(BCD) = dest(BCD) + C

Emulation DADD(.B or .W) #0,dest

Example:

```
DADC.B R7,R8            ; add two decimals in R7 to R8
JC     BAR              ; If result >99, jump to BAR
```

DEC (.B or .W) dest Decrement destination

Description: The destination is decremented by 1.

Operation: dest = dest-1

Emulation SUB(.B or .W) #1,dest

Example:

```
      MOV  #4,R7        ; initialize counter for divide by 32
FOO   RRA  R5           ; Divide R5 by 2
      DEC  R7           ; Check for last divide
      JZ   BAR          ; If final sample, jump to label BAR
      JMP  FOO          ; else, loop to FOO
```

DECD (.B or .W) dest Double decrement destination

Description: The destination is decremented by 2.

Operation: dest = dest-2

Emulation SUB(.B or .W) #2,dest

Example:

```
    DECD    BAR         ; Decrement memory location BAR by 2
```

DINT Disable interrupts

Description: All interrupts are disabled.

Operation: GIE = 0

Emulation BIC #8,SR

Example:

```
    DINT                ; disables interrupts
```

EINT Enable interrupts

Description: All interrupts are enabled.

Operation: GIE = 1

Emulation BIS #8,SR

Example:

```
    EINT                ; enables interrupts
```

INC (.B or .W) dest Increment destination

Description: The destination is incremented by 1.

Operation: dest = dest+1

Emulation ADD(.B or .W) #1,dest

Example:

```
        MOV  #0,R7          ; initialize counter fo
FOO     ADD  BASE(R7),R5    ; Add new value to R5
        INC  R7             ; increment counter
        CMP  #8,R7          ; Check for last sample
        JZ   BAR            ; If final sample, jump to label BAR
        JMP  FOO            ; else, loop to FOO
```

INCD (.B or .W) dest   Double increment destination

Description: The destination is incremented by 2.
Operation: dest = dest+2
Emulation Add(.B or .W) #2,dest
Example:

```
    INCD    BAR         ; Decrement memory location BAR by 2
    JC      FOO         ; Jump to FOO if carry
```

INV (.B or .W) dest   Invert destination

Description: The bits in destination are inverted.
Operation: dest = NOT.dest
Emulation

```
    XOR   #0FFFFh,dest  ; word operation
    XOR   #0FFh,dest    ; byte operation
```

Example:

```
    INV     R6          ; inverts bits in R6
```

NOP   No operation

Description: No operation is performed. Typically, this instruction is used to fill time for code synchronization purposes.
Operation: None
Emulation MOV #0,R3

POP (.B or .W) dest   Pop stack to destination

Description: The value at TOS is moved to destination, and the stack pointer is incremented accordingly.
Operation: dest = @SP, SP=SP+2
Emulation MOV(.B or .W) @SP+,dest

Example:

```
    POP     R6          ; restores R6 from stack
    POP     SR          ; restores status register from stack
```

RET Return from subroutine

Description: The complement of the CALL mnemonic, RET pops the PC address from the stack

Operation: PC = @SP, SP=SP+2

Emulation MOV @SP+,PC

Example:

```
RET                     ; returns from subroutine call
```

RLA (.B or .W) dest Roll left arithmetically

Description: All bits in the destination are shifted left one bit location. The LSB is reset, and the MSB is shifted into the carry flag.

Operation:
dest (MSB)=>C
dest (MSB-1)=>dest(MSB)
:
:
dest (LSB)=>dest(LSB+1)
dest (LSB) = 0

Emulation ADD(.B or .W) dest,dest

Example:

```
        MOV   #4,R7     ; initialize counter for multiply by 32
FOO     RLA   R5        ; multiply R5 by 2
        DEC   R7        ; Check for last multiply
        JZ    BAR       ; If final multiply,jump to BAR
        JMP   FOO
```

RLC (.B or .W) dest   Roll left through carry

Description: All bits in the destination are shifted left one bit location. The LSB is loaded with carry, and the MSB is shifted into the carry flag.

Operation:
```
dest (MSB)=>C
dest (MSB-1)=>dest(MSB)
      :
      :
dest (LSB)=>dest(LSB+1)
dest (LSB) = C
```

Emulation   ADDC(.B or .W)   dest,dest

Example:
```
BAR  RLC  R15       ; This set of commands shifts
     RLC  R14       ; 64 bits of data through the
     RLC  R13       ; processor registers and into the
     RLC  &P4OUT    ; Port 4 output register
```

SBC (.B or .W) dest   Subtract borrow from destination

Description: The borrow is subtracted from the destination. Borrow is defined as the complement of the carry flag.

Operation: dest = dest– 1 + C

Emulation   SUBC (.B or .W)   #0,dest

Example:
```
     SBC  R12       ; Subtract C from R12
     JN   BAR       ; If R12 = 0 and C = 1, jump to BAR
```

SETC Set carry flag

Description: Carry flag is set.

Operation: C = 1

Emulation BIS #1,SR

Example:

```
    SETC                    ; sets carry flag
```

SETN Set negative flag

Description: Negative flag is ret.

Operation: N = 1

Emulation BIS #4,SR

Example:

```
    SETN                    ; sets negative flag
```

SETZ Set zero flag

Description: Zero flag is set.

Operation: Z = 1

Emulation BIS #2,SR

Example:

```
    SETZ                    ; sets zero flag
```

TST (.B or .W) dest Test destination

Description: Destination is tested for a zero condition

Operation: dest-0

Emulation CMP (.B or .W) #0,dest

Example:

```
    TST     R12             ; tests R12
    JZ      BAR             ; if R12=0, jump to BAR
```

Language Selection

While it is helpful for understanding the functionality of the device, it is not unlikely that you, the reader, might never actually use this instruction set directly. Very few companies develop in assembly language anymore. Because of the effects of Moore's Law and increasingly complex designs, development time has become the critical resource. Generating 4k or more of tight, stable assembly code simply takes too long. C language has become the tool of choice for many designs, since it provides the best combination of direct access to hardware and rapid, efficient development. Additionally, most of today's commercially available platform-dependent compilers will develop code which is just as tight as an assembly programmer. (I know that I have just raised the ire of hundreds of assembly programmers out there, but the days of C compilers that produce bloated, slow code are nearing their end.)

When selecting a language for development, you should ask yourself the following questions:

1. How large is the project? If your objective only requires 50 instructions, programming and debugging in assembly can be done in an afternoon. If you are developing a 10,000 instruction control system, a structured language makes more sense.
2. Is portability an issue? If you plan on reusing code, in whole or part, on other platforms, you should probably rule out assembly immediately.
3. What resources do you have available? Most hobbyists aren't going to shell out the $2000 or so for a full-blown ANSI C compiler. There are various freeware compilers for the MSP430 out there, of unknown quality and completeness. If you are interested, post a question about these compilers on comp.arch.embedded. There are usually a couple of finished compilers, and one or two more "in progress."

4. Where is your comfort zone? You should be comfortable in the language you select, and understand its features. For example, use of Forth on microcontroller projects has been growing in popularity in recent years. I have never used the language myself, but the few Forth programmers I have spoken with are almost religious in their loyalty. I would never suggest that these developers should abandon their language in favor of C.

CHAPTER 10

Flash Memory

Few improvements in the past decade have aided in the development of embedded software as much as flash memory. Tasks that used to require expensive emulation systems can now be performed on the components themselves, for less than the cost of this book. External EEPROM is no longer necessary in many cases, as critical values can be saved to on-board flash memory. When bugs occur (and they will), we can simply reprogram, rather than replace. I cannot remember how I survived before flash memory.

Ever responsive to the market, TI has produced flash versions of most '430 devices. They are numbered as MSP430Fxxx, and are slightly more expensive than the ROM or OTP parts. Additionally, there is a low-cost flash emulation tool available from TI.

Flash Memory Structure

Flash memory is divided into segments in the '430 devices. Flash memory may be written one byte or word at a time, but must be erased in segments. Erased flash locations hold the value 0xFF(FF). These devices have the ability to "overprogram," that is, to reprogram the same location multiple times between erase cycles. You can always turn 1s into 0s, but you must erase the segment to change 0s back to 1s.

There are two different type of flash memory segments. The first is information memory, which is always two 128-byte segments, at locations

0x10FF-0x1080 (Segment A) and 0x107F-0x1000 (Segment B). These segments are intended for use as on-board EEPROM, although nothing prohibits you from executing code from this space. The second type of segment is more general. In 'Fxxx devices, all of the code space is divided into 512-byte segments, beginning at the top of memory. Segment 0 is located from 0xFFFF-0xFE00, Segment 1 is located from 0xFDFF-0xFC00, and so on.

Flash Memory Control Registers

- FCTL1, Flash Memory Control 1.
  Address: 0x0128h
  All unreserved bits are readable and writable.
  Structure:

Bit	R/W	R/W	R/W	R/W	R/W	R/W	R/W	R/W
Reset Value	0	0	0	0	0	0	0	0
Bit Position	15 (MSB)	14	13	12	11	10	9	8

Bit	SEGWRT	WRT	res	res	res	MEras	Erase	res
Reset Value	0	0	0	0	0	0	0	0
Bit Position	7	6	5	4	3	2	1	0 (LSB)

R/W : Read/Write
This byte is used for write security for the register. When the FCTL1 register is read, this byte will read as 0x96. When writing to this register, the value 0xA5 must be written to this byte. Any attempt to write to this register with a different high byte will generate a flash access violation interrupt. (This is one method by which your code can perform a soft reset).

SEGWRT : Segment Write
This variable is also referenced as BLKWRT in some TI literature. This bit is set high when writing an entire block to flash memory. When writing multiple flash blocks, this bit must be brought low, then high again, between blocks.

WRT : Write
This bit must be set high for a valid write operation. If a write operation is attempted when this bit is reset, an access violation interrupt is generated.

MEras : Mass Erase
When this bit is set, a mass erase is executed, by performing a dummy write operation into a segment. With the mass erase operation, all segments, from segment 0 to the targeted segment, inclusive, are erased.

Erase : Erase
When this bit is set, an erase is executed, by performing a dummy write operation into a segment. With the erase operation, only the targeted segment is erased.

- FCTL2, Flash Memory Control 2.
  Address: 0x012Ah
  All unreserved bits are readable and writable.
  Structure:

Bit	R/W	R/W	R/W	R/W	R/W	R/W	R/W	R/W
Reset Value	0	0	0	0	0	0	0	0
Bit Position	15 (MSB)	14	13	12	11	10	9	8

Bit	SSEL1	SSEL0	FN5	FN4	FN3	FN2	FN1	FN0
Reset Value	0	0	0	0	0	0	0	0
Bit Position	7	6	5	4	3	2	1	0 (LSB)

R/W : Read/Write
This byte is used for write security for the register. When the FCTL2 register is read, this byte will read as 0x96. When writing to this register, the value 0xA5 must be written to this byte. Any attempt to write to this register with a different high byte will generate a flash access violation interrupt. (This is one method by which your code can perform a soft reset).

SSEL : Clock Source Select
This variable determines the clock source for the flash timing generator.

SSEL=00 ACLK
SSEL=01 MCLK
SSEL=10 SMCLK
SSEL=11 SMCLK

FN : Clock Division Rate
The flash timing generator rate is divided by the value (FN+1)

- FCTL3, Flash Memory Control 3.
  Address: 0x012Ch
  All unreserved bits except WAIT are readable and writable.
  Structure:

Bit	R/W	R/W	R/W	R/W	R/W	R/W	R/W	R/W
Reset Value	0	0	0	0	0	0	0	0
Bit Position	15 (MSB)	14	13	12	11	10	9	8

Bit	res	res	EMEX	Lock	WAIT	ACIFG	KEYV	BUSY
Reset Value	0	0	0	0	0	0	0	0
Bit Position	7	6	5	4	3	2	1	0 (LSB)

R/W : Read/Write
This byte is used for write security for the register. When the FCTL3 register is read, this bytc will read as 0x96. When writing to this register, the value 0xA5 must be written to this byte. Any attempt to write to this register with a different high byte will generate a flash access violation interrupt. (This is one method by which your code can perform a soft reset).

EMEX : Emergency Exit
This is your trapdoor, for when the flash write locks itself in an eternal loop. A 1 written to this bit completely shuts down the flash memory controller, resets FCTL1, and resets itself.

Lock : Lock Bit
Setting this bit prevents writing or erasing of flash memory. This bit is fully software controllable.

WAIT	:	Wait bit When in block write mode, this bit indicates when the flash memory controller is ready for the next byte or word. A 0 indicates that programming is under way, and a 1 indicates that the memory is ready for the next write.
ACIFG	:	Access violation interrupt flag This flag is set when flash is improperly accessed. Along with this, an interrupt request is sent. In order for the interrupt to activate, the ACCVIE bit in IE 1 must also be set.
KEYV	:	Key Violated Bit This bit is set high when a write to FCTL1, 2, or 3 is attempted and the high byte was NOT set to 0xA5. This bit is not reset automatically. Setting this bit high will prompt a PUC.
BUSY	:	Busy Bit If this bit is 0, access to flash memory is possible. If this bit is 1, attempted access to flash will cause an access violation. This bit should be tested prior to any write or erase attempt. This bit will remain high if a block write function is underway.

Using Flash Memory

Erasure and writing of flash memory is a relatively straightforward process. All flash activities are internally timed, based on the Flash Timing Generator. The timing generator is initialized in FCTL2, and the speed of the generator is specified in the device datasheet. Once the timing generator is set up, its functionality is transparent to the developer and the firmware. I typically set it up to be about 400 kHz, and then forget about the thing. The erase and write processes described below assume that the timing generator is properly configured.

Segment and mass erase processes each require about 5,000 cycles of the Flash Timing Generator. The segment erase is a bit faster then the mass erase (about 10%). Erasing flash memory requires several sequential steps. They are:

1) Check the BUSY bit, in FCTL3.
2) Clear the Lock bit, in FCTL3.
3) Set Erase, or MEras, in FCTL1, depending on whether you are performing a segment erase or a mass erase.
4) Perform a dummy write to the segment to be erased. Any write, clear, or logical operation performed on any address in the correct segment will work.
5) Wait for the BUSY bit, in FCTL3, to go low.
6) Set the Lock bit, in FCTL3, to prevent accidental writes.

A single element (byte or word) write requires 33 cycles of the Flash Timing Generator. Writing to flash memory is a similar process. A single element is written using the following process:

1) Check the BUSY bit, in FCTL3.
2) Clear the Lock bit, in FCTL3.
3) Set the WRT bit, in FCTL1.
4) Write the element to the proper address, using a mov.b or mov.w instruction. This starts the timing generator.
5) Wait for the BUSY bit, in FCTL3, to go low.
6) Set the Lock bit, in FCTL3, to prevent accidental writes.

A block write is similar to a repetitive element write, the main difference being that the block write process requires about half as long as repeated single element writes would. Blocks are predefined 64-byte structures in memory. Each block begins at addresses of the form xxxx xxxx xx00 0000 (e.g 0xFE80 or 0xFCC0), and ends 64 bytes later, at an address of the form xxxx xxxx xx11 1111 (e.g. 0xFEBF or 0xFCFF). The block write process requires about 20 cycles of the Timing Generator per element, plus overhead of about 15 more cycles. The process is:

1) Check the BUSY bit, in FCTL3.
2) Clear the Lock bit, in FCTL3.
3) Set the WRT and BLKWRT bits in FCTL1.
4) Write element to proper address.
5) Loop until WAIT bit, in FCTL3, is set.
6) Repeat from step 4) until all elements have been written.
7) Clear the WRT and BLKWRT bits in FCTL1.
8) Wait for the BUSY bit, in FCTL3, to go low.
9) Set the Lock bit, in FCTL3, to prevent accidental writes.

Remember that there is a minimum time between sequential block writes, specified in the datasheet, that must be respected. It is generally a few milliseconds, and a block write typically requires about 25 milliseconds.

Security Fuse

The MSP430 flash devices contain a security fuse, which is settable through the JTAG port. Setting of this fuse is described in the TI application note "Programming a Flash-Based MSP430 Using the JTAG Interface," number SLAA149. The process is automated through most compiler/programmer tools. The important thing to note is that the security fuse is a one-time only burn. This means that, once the fuse has been set, further flash writes and erasures are impossible. You have just turned the flash device into an OTP device. Burning the fuse not only restricts flash programming, it prohibits further JTAG access.

Information Memory

Along with the regular flash memory blocks, there are two 128-byte segments, commonly referred to as information memory. Information memory is intended to function as on-chip EEPROM, and is similar to regular flash. The most notable differences are:

- Information memory segments are only 128 bytes in size, rather than the standard 512 bytes.
- Information memory is located at lower memory addresses, immediately following RAM, in the address space.
- A standard mass erase function will not erase the information memory segments. To include the information memory segments in a mass erase, set both the MEras and Erase bits in FCTL1.

I commonly use the information memory for storage of device performance parameters. This allows for editing of the parameters when necessary, but preserves the parameters through power loss. One consideration to keep in mind is that of programming cycles. If your application erases and rewrites information memory every few seconds, it will exceed the specified number of write cycles in a relatively short amount of time. Information memory, and flash in general, are intended for occasional or long-cycle time storage.

Flash Memory Code Examples

Code example 10.1 erases information memory segment A, located between addresses 0x1080 and 0x10FF. It is assumed that MCLK is configured at the maximum DCO rate. Information memory block A is selected by the address defined in temp_local_pointer, and this value can be changed to erase any valid flash block.

Code Example 10.1: Block Erase Function

```
void EraseInformationMemory_A(void)
    {
    char *temp_local_pointer = (char *) 0x10FF;

    FCTL2 = FWKEY + FSSEL_1 + 12; //Set Up Timing Generator
                                  //for MCLK and a clock
                                  //divisor of 12
    FCTL3 = FWKEY;                //Make Certain that LOCK
                                  //is cleared

    FCTL1 = FWKEY + ERASE;
    *temp_local_pointer = 0xFF;
    while (FCTL3 & 0x0001);    //Wait for BUSY flag to clear
    return ;

    }//End EraseInformationMemory_A Function
```

Code example 10.2 performs a looped element write to information memory block A, which begins at address 0x1080. As in the previous example, the timing generator parameters assume use of the DCO at maximum rate for ACLK. It is important to note that this is not necessarily the most efficient way to perform this function. The segment write function, as described earlier in the chapter, requires far fewer clock cycles to complete. I tend to prefer this method, however, as it is much more generalized. The looped element method allows for writes over block boundaries.

Code Example 10.2: Looped Element Flash Write Example

```
void WriteInformationMemory_A(unsigned int length)
    {
    int *temp_local_pointer = (int*) 0x1080;
    unsigned int LoopIndex=0;

    FCTL2 = FWKEY + FSSEL_1 + 12; // Set Up Timing Generator
    FCTL3 = FWKEY;                // Make Certain that LOCK is
                                  // cleared

    for (LoopIndex=0;LoopIndex<length;LoopIndex++)
        {
        FCTL1 = FWKEY + WRT;
        *parameter_pointer = Element_Array[LoopIndex];
        while (FCTL3 & BUSY);
        parameter_pointer+=2;
        }//End For

    return;
    }//End WriteInformationMemory_A function
```

Code example 10.3 is something of a corner case. It literally erases all of flash memory (in the 'F149, which has address space down to 0x1100). There are very few practical applications where this is useful, and would need to be run from RAM space, but it is useful as an example for several reasons. It serves to illustrate the mass erase function. It also illustrates a critical error condition that your code should strive to avoid. Note that this is nearly identical to example 10.1, which erases a single block only.

Code Example 10.3: Erase Everything

```
void EraseEverything(void)
    {
    char *temp_local_pointer = (char *) 0x110F;

    FCTL2 = FWKEY + FSSEL_1 + 12; //Set Up Timing Generator for
                                  //MCLK and a clock divisor of 12
    FCTL3 = FWKEY;                //Make Certain that LOCK is
                                  //cleared


    FCTL1 = FWKEY + MERAS;
    *temp_local_pointer = 0xFF;
    while (FCTL3 & 0x0001);    //Wait for BUSY flag to clear
    return ;

    }//End EraseEverything Function
```

Bootstrap Loader

The term "bootstrap loader" has historically meant a particular device or piece of ROM code, which loads specific instructions to be executed into a device on powerup, for initialization purposes. That is not the case with the '430 flash devices. The bootstrap loader is an external interface, similar to the JTAG, which may be used to program flash memory. Like the JTAG, TI has gone to great lengths to document the implementation and use of the bootstrap. If you intend to use the bootstrap, hit the TI website, and download Application Report SLAA089A, "Features of the MSP430 Bootstrap Loader". Some of the high points are:

- The bootstrap loader code is stored in a special section of ROM, which is untouchable by other applications, so you need not worry about accidentally overwriting it.
- The loader is triggered by cycling the TCK pin on the JTAG port low, then high, then low again, and bringing reset high. Shortly after the reset, bring the TCK pin high again, and the bootstrap loader will start.
- The UART communication protocol is used by the bootstrap loader, at a fixed data rate of 9600 baud. A proprietary TI data structure protocol is used.
- The bootstrap loader can perform essentially the same functions as the JTAG interface, with the exception of the security fuse. The bootstrap cannot program the security fuse, while the JTAG can.

Downloadable Firmware

One increasingly common feature of embedded devices is that of downloadable firmware. It is useful for both correction of bugs and addition of design features after product deployment. If possible, the easiest, fastest, and most reliable method for doing this is with a hard connection, either JTAG or bootstrap loader. Often, however, particularly in wireless devices, the download process must be controlled by software.

There are three common approaches to the downloadable firmware problem. The first is the dual code partition solution. In this approach, the device is selected to have more than twice as much code space as is actually necessary. The initial code is run entirely in the top half of the code space. When it becomes time to download a new code set, it is read in, and written to the second half of the device code space. The vector table is then re-written to match the new code, and a soft reset is performed. This approach is straightforward, simple, and easy to implement. It also allows you to replace everything. It is, however, expensive, because it requires a more expensive device than you really need.

The second is the brute force approach. It is a piecewise implementation of the first approach. The code is downloaded, one block at a time, and written directly to an available flash block, where it is checked for error. If it passes the error check, it is then copied to its final location. This process is performed sequentially until the load is complete. It requires less memory than the first approach, but prohibits loading of certain functions, including the communications block.

The third approach, downloadable programmer, is popular with larger platforms. In this approach, a small program, which controls the firmware download process, is loaded into RAM, and then executed from RAM. This is possible with von Neumann devices like the '430, and the larger devices have enough RAM space to make it possible. The advantage of this approach is that it requires almost no added code to be implemented in the device. The disadvantage is reliability: if there is a power loss or inadvertent reset (like a WDT timeout), the device may become unrecoverable. However, if you can find a suitable workaround, this approach really is clean.

CHAPTER 11

Developer's Toolbox

This chapter is intended to offer code modules for some common functions. Examples in this chapter are offered in C, rather than assembly language, for clarity and usability. The designs and code in this chapter were designed, generally, for a '1xx device. They will, however, operate properly on '3xx and '4xx devices with little or no modification.

Real-Time Clocks

Many different applications require tracking of real-world time, for monitoring, triggering or marking of events. Typically, time is tracked by one of two methods. The first is a simple count of seconds elapsed since some predefined mark. The second method counts seconds, minutes, hours, and days. Each method has its advantages, and both are described here.

Before delving into the code, configuration and setup conditions must be discussed. The design assumes there is a 32.768 kHz crystal connected at XIN/XOUT, and uses Timer A to count seconds based on that input signal. There is nothing special about these selections, and you should be able to modify this example to perform the clock function with Timer B, and/or other clock sources and speeds.

Generally speaking, initialization needs to be performed on both the Basic Clock Module and the Timer A controls. However, the BCM initialization is trivial. As long as DIVA is set to zero (i.e., ACLK is divided by 1) and the XTS bit is cleared, the clock controls are ready. Timer A will require several more steps, however. They include:

- Select ACLK as the timer source.
- Enable the proper interrupt.
- Set up Capture/Compare unit 0 control, TACCTL0.
- Initialize the Capture/Compare unit 0 register, TACCR0.
- Put Timer A into Compare mode.

These basic initialization steps are implemented with the following statements:

Code Listing 11.1: Timer A Initialization

```
TACTL = MC_0 + ID_0 + TASSEL_1 + TACLR + TAIE;
                          //Leave Timer in Stop Mode During
                          //Initialization
                          //Do Not Divide Input Timer
                          //Select ACLK as the Timer A Source.
                          //Begin with Timer A Cleared
                          //Enable Timer A Interrupt
TACCTL0 = CCIE;           //Enable Capture/Compare Interrupt
TACCR0 = 32767;           //Set TimerA Upper Count Limit.
                          //32768 states, including 0.
TACTL |= MC_1;            //Turn on Timer A Using Bitwise OR
```

Some notes and thoughts on Listing 11.1:

- As with all code in this book, standard TI register names and bit definitions are used. The include file which maps registers to these names is available at the TI website, or on the development kit CD-ROM.
- This code should be included as part of the initialization set. The last line, which starts Timer A, should not necessarily be with the rest of this code. Rather, this function should be performed when setup is complete, and the code is ready to begin processing interrupts.

- Since we are using Capture/Compare 0, the processing will be performed from the lower priority Timer A interrupt, at location 0xFFEC. If Capture/Compare 1 or 2 are used, the interrupt vectors from location 0xFFEA.
- The design here does not divide the input timer. However, if execution cycles or current consumption are tight, this can be divided, and the code can count in 2, 4 or 8 second increments.
- Along with proper Timer and Clock control, time variables need to be defined. Since each of the two clock approaches keeps different variables, they will be defined separately.

UTC Time

One very common approach to time tracking is to use UTC time, which is a 32- bit count, representing seconds since midnight, Greenwich Mean Time, on January 1, 1970. There are also various derivatives of this, using different time zero values. This approach is very simple, requiring only a single increment operation when the interrupt occurs. It is also a very handy method when it is necessary to determine the elapsed time between two events, requiring a single subtraction operation to find this difference. The third advantage of this method is that it is completely indifferent to time zones, leap years, and Daylight Savings Time.

Code Listing 11.2: UTC Real Time Clock

```
//Variable Declaration
unsigned long    UTC_Count=0;    //This declaration belongs at the
                                 //start of code.

//Timer A Interrupt Routine
{
UTC_Count++;                     //Increment Real Time Clock
return;                          //Return from interrupt.
}
```

Some notes and thoughts on Listing 11.2:

- This approach to the Real Time Clock really is as simple as it looks. There is a single variable that is incremented every second, and that variable is used to mark time.
- As you probably noticed, the time variable, UTC_Count, is initialized to zero. It is quite likely that you will need a communications method and subroutine to set the time. If your application is concerned with what time it is in the real world, it will need to be initialized to a different value, and periodically corrected, to account for oscillator errors.
- I have written the body of the Interrupt Service Routine, but omitted the function declaration. This is because there are different ways to structure ISRs, depending on your compiler and style. Chapter 2, *Resets and Interrupts*, offers one very common structure for these ISRs.

Calendar Time

Another common approach is to count seconds, minutes, hours, days, and so on. The code more closely matches common time measurements, but is more complicated to implement, and to correct.

Code Listing 11.3: Calendar Based Real-Time Clock

```
//Variable Declaration
unsigned char   seconds=0;        //These declarations belong at the
                                  //start of code.
unsigned char   minutes=0;
unsigned char   hours=0;
unsigned char   day_of_week=1;    //Sun=1, Mon=2, Tue=3,
                                  //Wed=4, Thur=5, Fri=6, Sat=7
unsigned char   day_of_month=1;
unsigned char   month=1;          //Jan=1, Feb=2, etc.
unsigned int    year=2000;
```

```
//Timer A Interrupt Routine
{
    seconds++;              //Increment Real Time Clock
    if (seconds >= 60)
        {
        seconds=0;
        minutes++;
        if (minutes >=60)
            {
            minutes=0;
            hours++;
            if (hours >= 24)
                {
                hours=0;
                day_of_week++;
                if (day_of_week >= 8) day_of_week = 1;
                day_of_month++;
                if ((month=1 && day_of_month >=32) ||
                    (month=2 && day_of_month>=30 && !(year%4)) ||
                    (month=2 && day_of_month>=29 && year%4) ||
                    (month=3 && day_of_month >=32) ||
                    (month=4 && day_of_month >=31) ||
                    (month=5 && day_of_month >=32) ||
                    (month=6 && day_of_month >=31) ||
                    (month=7 && day_of_month >=32) ||
                    (month=8 && day_of_month >=32) ||
                    (month=9 && day_of_month >=31) ||
                    (month=10 && day_of_month >=32) ||
                    (month=11 && day_of_month >=31) ||
                    (month=12 && day_of_month >=32))
                    {
                    day_of_month=1;
                    month++;
                    if (month >=13)
```

```
                          {
                          month=1;
                          year++;
                          }//End month if statement
                      }//End day of month if statement
                  }//End hours if statement
              }//End minutes if statement
          }//End seconds if statement
    return;                                //Return from interrupt.
}
```

Some notes and thoughts on Listing 11.3:

- The difference in complexity between this method and the UTC approach jumps off the page, doesn't it? Five levels of if statements, and look at the decision tree required to decide if it is the end of the month. This method is equally complex on which to perform periodic corrections. Keep in mind, this method will need to correct for Daylight Savings Time.
- To avoid this complexity, it is a common approach to track the time by counting seconds, as in the UTC approach, and perform a conversion to hours, minutes, and seconds when required. This approach is just as difficult to implement, but easier to update, and requires much less processing in the ISR.
- Many applications will require only a subset of this design. If your code is only being designed to take hourly measurements, this can be implemented without the day of week, day of month, month and year fields, making the code considerably simpler.

Using Time

Now that we are measuring time in software, how to use it? When seconds (or minutes, or hours) tick off, it is common to check to see if it is time to perform a timed function. I prefer to keep as much functionality outside of the ISR as possible, and the best way to do that is to set a flag in the ISR, and come back and service it later. In code listing 11.4, the UTC method is modified to do this.

Code Listing 11.4: UTC Real-Time Clock with Event Checking

```
    //Variable Declaration
    unsigned long UTC_Count=0;          //This declaration belongs at the
                                        //start of code.
    unsigned int   Flags=0x0000;        //General bitflag field. The most
                                        //significant bit is the timer
                                        //increment flag


    //Somewhere in the main loop......
    if (Flags & 0x8000)
        {
        Flags -= 0x8000;
        checkForTimedEvent();           //Generic call to subroutine.
        }//End Time-Based event if

    //Timer A Interrupt Routine
    {
    UTC_Count++;                        //Increment Real Time Clock
    Flags |=0x8000;                     //Set Timer Increment Flag
    return;                             //Return from interrupt.
}
```

Some notes and thoughts on Listing 11.4:

- This example checks each second. If you only need to check every minute, hour, or fortnight, you can use another counter to count, incremented with UTC_Count and reset each 60 seconds (or any other length) to trigger the flag.

- Make certain that the time required to reach the check of the flag in the main loop PLUS the time required for any processing that is performed within checkForTimedEvent() totals less than 1 second at your selected processor speed, so the function is complete before it is called again. Otherwise, the stack will fill up RAM, and your code will crash.

Error Detection

Any time your application is required to transfer more than a few bytes of data to or from the outside world, it is a good idea to use some basic error control. This control can be divided into two groups: error detection and error correction.

Error detection is simply meant to detect whether data is valid, with no means of correcting invalid data. Most methods involve generating a checksum from the data, and appending it to the data block. We will briefly examine the most common of these methods, the Cyclic Redundancy Check, or CRC.

This method generates a CRC checksum. The checksum can theoretically be any power of 2 in length, but for practical purposes, is almost always 16 or 32 bits. We will use the 16-bit case (commonly referred to as CRC-16). In order to generate the checksum, a CRC polynomial is required. The term polynomial can be a bit confusing, because it is simply a bitmask in the application. For the CRC-16, the polynomial is of degree 16, with the non-zero terms of the polynomials corresponding to bits in the mask. For example, the polynomial $x^{16}+x^{12}+x^5+1$ equates to a bitmask of 0x1021. (In CRC-16 polynomials, the x^{16}, or 17th, term is always high, but it is not represented in the bitmask.) There are different standardized polynomials, and we will use the above example 0x1021, which is the CCITT standard.

CRC coding is based on modulo-2 division. The basic concept is that the message block, modulo-2 divided by the polynomial, yields the CRC-checksum. The same message, this time with the CRC checksum included at the end, modulo-2 divided by the polynomial, will result in 0. This is advantageous in that you can use the same code to generate and check CRC correctness.

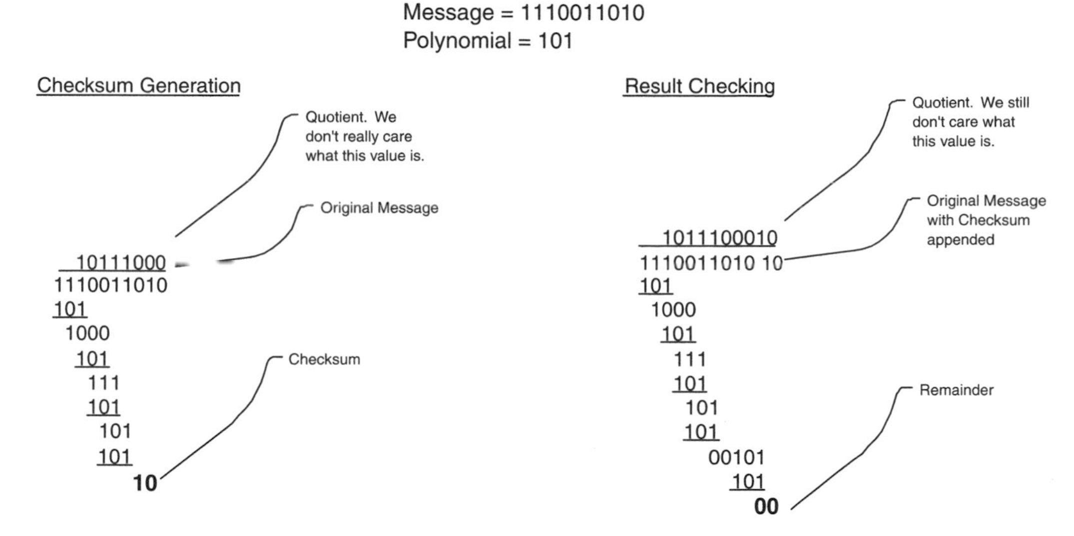

Figure 11.1: CRC Calculation Example

There are two standard methods for CRC implementation. The common table method is fast, but requires more code space. The loop algorithm requires much less ROM, but is slower by an order of magnitude. There are many sources for algorithms and code for both of these. We will look at the looping algorithm.

The looping algorithm is based on hardware implementations, which calculate the checksum using a Linear Feedback Shift Register (LFSR) (see Figure 11.2). The register is pre-loaded with the first data word of the message, and then the message is shifted into the register, with the XOR function being performed (XOR is equivalent to modulo-2 division). The flowchart for the algorithm is shown in Figure 11.3.

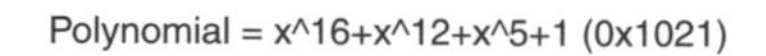

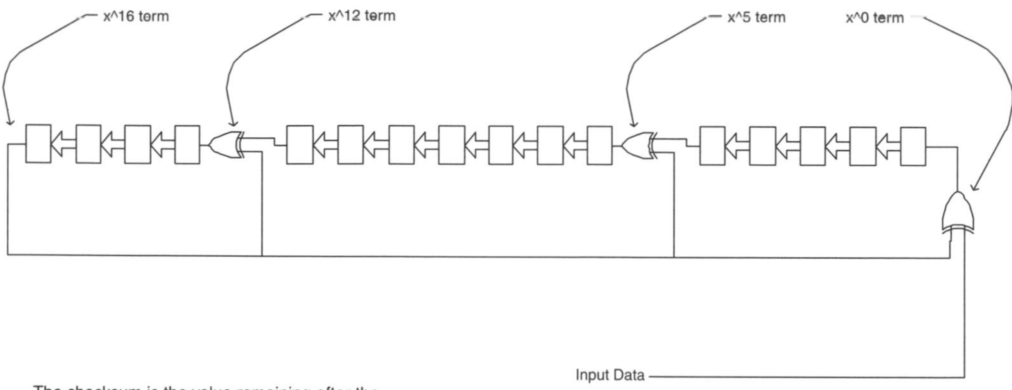

Figure 11.2: CRC Loop Example

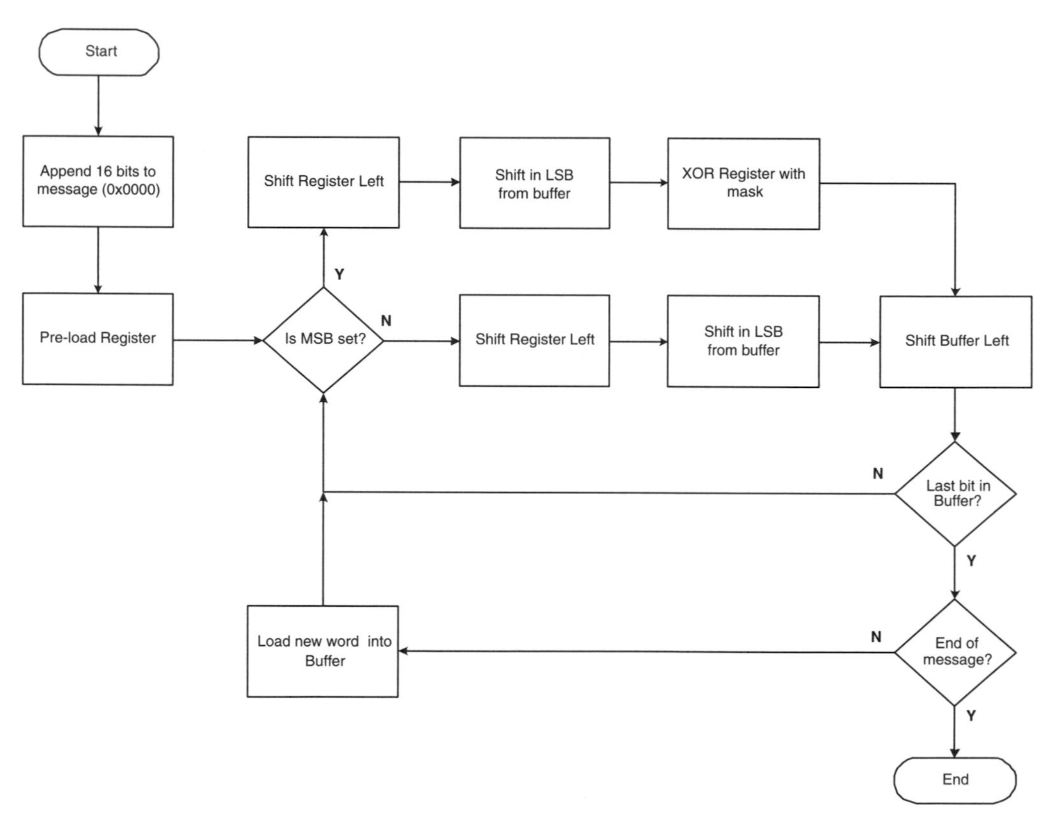

Figure 11.3: CRC Example Flowchart

Code Listing 11.5: Loop Algorithm for CRC Computation

```
unsigned int  Compute_CRC(unsigned int Data [], unsigned int Length)
{
unsigned int CRC_Working=Data[0];
unsigned int Buffer=Data[1];
unsigned int Counter=1;
unsigned int LoopIndex=0;

while (Counter<Length)
    {
        for (LoopIndex=0;LoopIndex<16;LoopIndex++)
            {
            if (CRC_Working >=0x8000)
                {
                CRC_Working <<= 1;   //Shift CRC_working left
                if (Buffer>0x8000) CRC_Working++;
                                            //If MSB of buffer is high, it will
                                            //shift to LSB of Working
                CRC_Working = CRC_Working ^0x1021;
                                            //Apply XOR Mask
                }//End if
            else
            {
            CRC_Working <<= 1;         //Shift CRC_working left
            if (Buffer>0x8000) CRC_Working++;
                                            //If MSB of buffer is high, it will
                                            //shift to LSB of Working
            }//End Else
            Buffer<<=1;                //Shift Buffer left
            }//End For
        Counter++;
        Buffer=Data[Counter];
        }//End While
    return CRC_Working;
    }//End Function Compute_CRC
```

Code listing 11.5 provides a function which, when called, will compute the CRC-16 value for an array of 16 bit values. This is efficient for the MSP430, as it uses only the device's native data size for all values. It returns a value, which represents the checksum when used for generation, and should be zero for checking. If the checksum is non-zero when checking, you know an error has occurred. The code assumes that the message has already been properly appended.

The CRC is a very efficient way of detecting errors. If implemented properly, it will positively detect all single bit errors, double bit errors, and burst errors (where a sequential group of bits has been corrupted). The computational overhead is limited, and it requires only 16 bits to be added to the message. The downside is that it will only tell you whether or not an error has occurred, but not how to fix the error. If your application receives a corrupted message, according to the checksum, there is little that can be done other than discard the message. To solve this problem, there are many varying error correction schemes in wide publication. Unlike detection, where the use of the CRC is the dominant method, there is no one common correction scheme. There are entire texts written on the subject, describing the tradeoffs in data overhead, computability, and correctibility. For this reason, we will not attempt to tackle the subject here.

D/A Conversion: Pulse Width Modulation

As with many commercially available microcontrollers, the MSP430 has available a series of analog-to-digital converters (which are described in Chapter 6). The inverse operation of generating an analog level based on a digital value is not directly supported by the '430 hardware. It is, however, supported in the Timer A (and Timer B, where available) structure, which makes for easy implementation of a Pulse Width Modulator.

The PWM is not useful in all applications. If your design requires a very precise and stable analog level, you should look into a separate D/A device. However, for many applications, particularly control of power devices, this is a very easy and reliable method. It should be noted that the use of a PWM is not, strictly speaking, analog level generation (although a level could be generated with sufficient filtering). It is merely the rapid switching of a digital signal, which is used to turn a load device on and off rapidly enough that the effect is that of an analog control source. A perfect example is that of motor control (which happens to be the most common use of the PWM). If the controller is switching the motor on and off hundreds of times every second, the motor will behave as if the power source is being controlled, resulting in a motor speed which is proportional to the duty cycle of the PWM signal (see Figure 11.4), with very little variation.

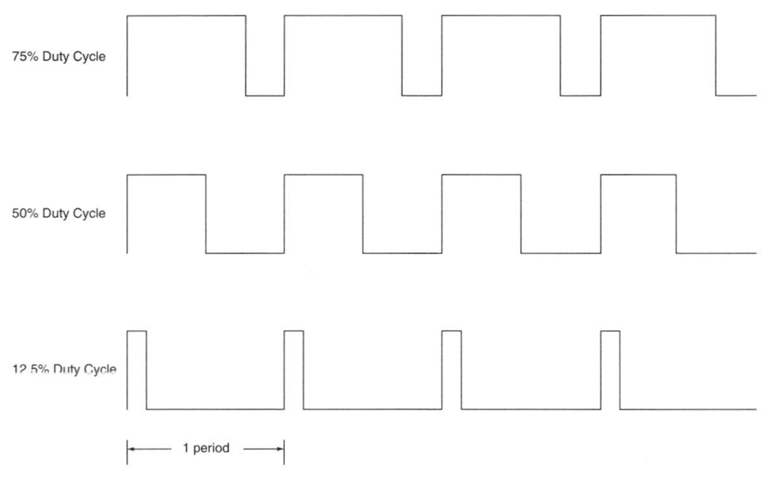

Figure 11.4: Pulse Width Modulation Sample Outputs

One very nice feature of the MSP430 family is that the PWM can be implemented fully in the Timer units. Your code can set up parameters, turn the timer on, and essentially forget about it until it is time to turn it off. While operating, the PWM requires no CPU cycles. There are four different PWM modes, two of which have "toggle" states, and the other two of which are strict set/reset implementations. (These are described in Chapter 4.) My personal preference is for a strict set/reset implementation, which is shown in Code Listing 11.6. (This implementation is not necessarily any better or worse than the toggle implementation, it is merely the one with which I am comfortable.)

Code Listing 11.6: PWM setup for MSP430F149

```
P1SEL |= 0x04;              //Output on p1.2
P1DIR |= 0x04;

TACTL |= TASSEL0;           //Use ACLK
CCR0 = 32767;               //Max value
CCR1 = 8198;                //75% duty cycle in this example (it would
                            //be 25% using OUTMOD_7)

CCTL1 = OUTMOD_3;           //Output Mode 3 is Set on CCR1, clear
                            //on CCR0
TACTL |= MC_1;              //Turn On Timer
```

As the code suggests, this is almost trivially simple to implement in '430 devices. It has been my experience that the hardware design has the potential to create more headaches than software on PWM implementations. In some cases, the various frequency components of the PWM signal can wreak havoc on other circuits. This, however, is a topic for another book.

Sliding Correlators

One common application performed in small embedded systems is recognition of a specific pattern in a sequence of bits. This is most commonly used in communications systems, when looking for identification tags. The simplest of sliding correlators shifts bits into a register, comparing the expected value to the shift register value with each bit shift. However, this is insufficient for many applications. Often, it is preferred to, on each new bit, determine how much correlation exists between the register and the expected mask. This is accomplished by XORing the shift register and expected mask together, and counting the number of bits that differ. A zero equates to an exact match, while a 16 means all bits are different. Typically, a return value of 0, 1, or 2 is necessary for any useful degree of correlation. Code Listing 11.7 is a function which can be used to compute correlation.

Code Listing 11.7: Sliding Correlator Comparison Function

```
unsigned int computeCorellationDistance(unsigned int Value)
    {
    unsigned int Mask=0x1F1F;          //Mask is the value we are checking
                                       //correlation to. This is an arbitrary
                                       //assignment in this example.
    unsigned int LocalBuffer=0;        //Temporary Local Value
    unsigned int LoopIndex=0;          //Loop Index
    unsigned int Counter=0;

    LocalBuffer=Value^Mask;            //Set upDifference Register using
                                       //XOR function


    for (LoopIndex=0;LoopIndex<16;LoopIndex++)
                                       //Count the bits that are different
    {
        if (LocalBuffer>=0x8000) Counter++;
                                       //Check MSB of LocalBuffer. If 1,
                                       //increment count
        LocalBuffer<<=1;
    }//End For

    return Counter;
}//End function computeCorellationDistance
```

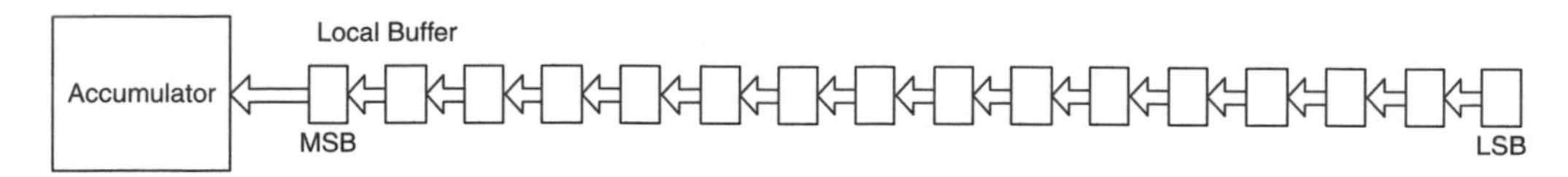

Figure 11.5: Sliding Correlator Construction

CHAPTER 12

Low-power Design

The rapid growth in the number of battery-operated devices has brought low-power design to the forefront of embedded systems development. The MSP430 is among the leading controllers in this regard, with typical room temperature consumption in the neighborhood of 1 mW/MIPS or less.

MSP430 Power Consumption Characteristics

As mentioned above, among the most salient features of the MSP430 is its very low power consumption. When you look at the first page of the datasheet, the sub-milliamp consumption values jump off the page. Being on the first page of the datasheet, these numbers tend to run on the optimistic side. As they say, your mileage may vary. However, if you are smart in your design, you could approach these values. Current consumption varies with clock speed, temperature, supply voltage, peripheral selection, input/output usage, memory type, wind speed, phase of the moon, and any other factor you might possibly imagine. Some general rules of thumb for predicting current consumption of your design:

- The datasheet provides several equations for active mode power consumption, that assume linearity over frequency and supply voltage. I have found these to be useful for a first-order approximation, but not sufficient for detailed current profiling. If you need to know, to a high order of accuracy, how much current your design is drawing, build a prototype and measure the consumption.

- Current increases with clock frequency, roughly linearly.
- Current increases with supply voltage. I have found this to be less linear than the datasheet would suggest.
- There is a relationship between supply voltage and maximum frequency. There is a graph describing this in the datasheet. The short version is that you can run at about 4 MHz at 2.2 VDC, but need to be at 3.6 VDC to operate at 8MHz. (I regularly exceed these limits on the bench, but wouldn't recommend it in production devices.)
- More active peripherals means more current consumption, but a Timer unit running at 8 MHz draws considerably less current than the CPU running at 8 MHz

MSP430 Low-power Modes

The '430 series devices have the option of shutting off the processor portion of the device, by using the CPUOff bit in the Status Register. The processor is then awakened by an interrupt. Further functionality in the CPU portion can be turned on and off via the SCG1, SCG0, and OscOff bits, which also reside in the SR. TI has, in the literature, defined five low-power modes. The bits and functionalities are shown in Table 12.1.

Table 12.1: Low-power Mode Summary

Mode	OscOff	SCG1	SCG0	SMCLK	ACLK	DCO	DCO Generator
LPM0	0	0	0	On	On	On	On
LPM1	0	0	1	On	On	On	Enabled if used by peripherals
LPM2	0	1	0	Off	On	Off	On
LPM3	0	1	1	Off	On	Off	Off
LPM4	1	1	1	Off	Off	Off	Off

When considering use of one of the low-power modes, the developer must remember several things:

- The CPU and MCLK are off for all low-power modes.
- The CPU is reactivated by an interrupt. Make certain that the GIE bit is set, or the wake-up will never occur.
- The interrupt can be either internal or external. If you are using one of the peripherals to generate the interrupt, it has the potential to use several orders of magnitude more current than the CPU in the low-power mode.
- As you might have noticed, LPM modes 0 and 1 are almost identical. If the DCO is being used, they will result in nearly identical current consumption.
- LPM modes 2 and 3 seem very similar, as well. This, however, is NOT the case. LPM2 leaves the DCO generator on, although the DCO oscillator is off. In these modes, ACLK is the only thing running in the CPU, and it takes roughly one-tenth as much current as the DCO generator. As a result, LPM2 uses about ten times as much current as LPM3 in the CPU.
- LPM4 turns everything off. It is common to see this listed as "RAM Retention Mode" in some literature. When in this mode, the device uses a mere trickle of current (around 100 pA typically). This mode is used for externally generated interrupts only, as no clocks are active and available for peripherals.
- The Status Register is pushed to the stack on interrupt, and popped back when the reti instruction is issued. Since the bits that determine the low-power mode are all part of the SR, the device will automatically reenter the mode upon completion of the ISR. This can be avoided by manually manipulating the SR on the stack. This is easy enough when developing in assembly language, but can be a bit trickier in C. In instances when I have used these modes, the code that I wrote performed all processing from within the ISR, rather than from within main(). At first, it seems a bit dubious, but can actually work out pretty well.

Periodic Interrupts and Low-Power Design

Perhaps the most common use of the low-power modes is for periodic processing based on a timer interrupt. The main loop is written so that startup housekeeping is performed, then the selected LPM is entered. Processing is then performed within the ISR. At the end of the ISR, the reti instruction restores the processor to the LPM. This is a great way to save power consumption when the application is required to make periodic decisions. Some hints and considerations:

- Select your power mode carefully. Obviously, LPM4 will not work for this, as all clock sources are shut down. After that, you must select the power mode based on your timer source. It is my personal preference to run off of ACLK and use LPM3, but not all designs will allow for that.
- Timing analysis is much easier in this mode. It essentially boils down to whether or not you can get through the ISR within your timer period. However, from a practical standpoint, time spent in the ISR is usually a very small fraction of the timer period. If your ISR time is the bulk of the timer period, the savings from the use of the LPM is minimal, and it probably makes more sense to use the ISR to set a flag and perform processing within the main loop.
- This method does not work well if your processing requires use of other interrupts. Since the interrupts are non-reentrant (see Chapter 3), your processing will not be able to rely on other interrupts from the timer ISR. There are two ways to work around this. The first is to manually set the GIE bit at the beginning of the ISR. I strongly recommend against this. It is an easy approach to implement, but introduces multiple possible error and crash conditions (again, see Chapter 3). The second approach is write the main loop so that, within an eternal loop, the main processing occurs, followed by the command to enter the LPM. The ISR then must manipulate the Status Register, on the stack, so that upon the return from interrupt, which vectors back to the next instruction, the processor is in active

mode, and loops through the code before looping back to the command to enter the LPM. This approach is a bit more complicated to implement, but is stable and predictable.

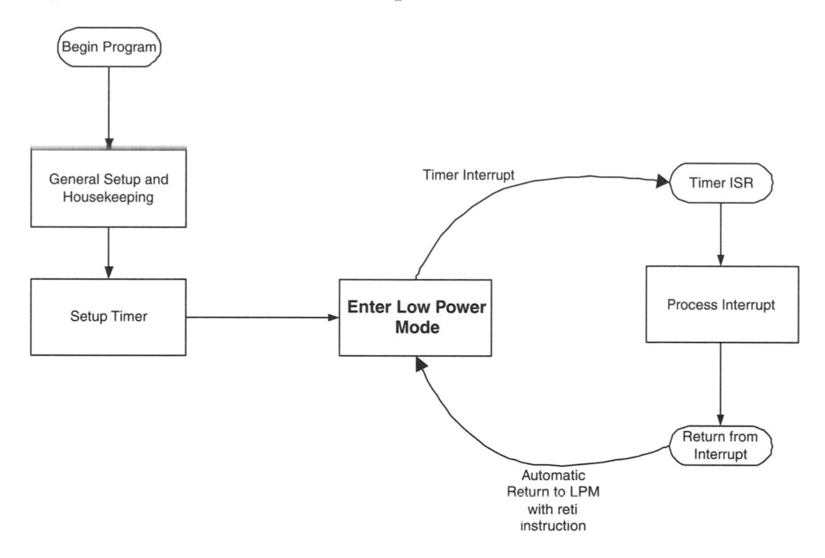

Figure 12.1: Periodic Interrupt Processing with Low-Power Mode

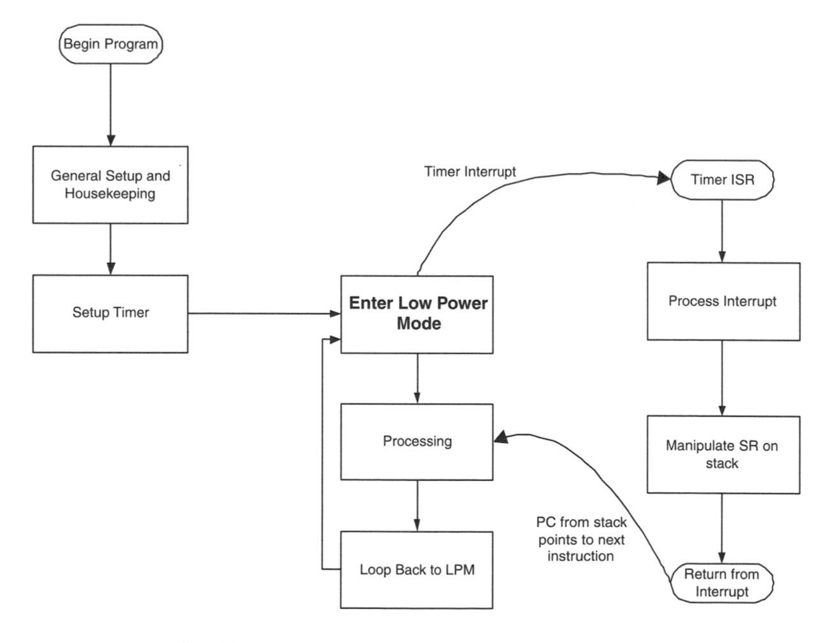

Figure 12.2: Periodic Main Loop Processing with Low-Power Mode

Low-power Design

So, now you have a very low-power microcontroller, and some idea of how to use it. So, what can you do if your circuit still uses too much power? Here are some rules for when you need to further reduce consumption:

1) Clock cycles are your enemy.

In CMOS devices, the rate of power consumption is based on changes in state on transistors, rather than the states themselves. The result of this is that doubling clock speeds means doubling power consumption. Remember, this doesn't only apply to the CPU. Use of a high-speed clock in Timer A will have the same effect.

The design approach to optimize this is obvious: Run your clocks as fast as you need to, but no faster. This sounds easier than it really is. Slow the clocks down just a little too much, and you could end up with a race condition on your hands. Even worse, the first race condition to show up as the clocks are slowed tends to be the irregular one, making it difficult to discover on the bench. As a result, this is an optimization which tends to create problems, even after testing, verification, and even after you have deployed thousands of units. Some rules of thumb for this optimization:

- Do plenty of timing analysis. Get a pencil, paper, and a calculator, and determine the absolute longest possible process loop, including any possible combination of ISRs. This is not as straightforward with the MSP430 as with other processors, since there is not a 1-to-1 relationship between instructions and clock cycles. However, you should be able to determine that you need to execute so many instructions in so much time, and clock speed comes from that analysis. Adding some buffer cycles by speeding the clock up by a few percentage points will hurt the power consumption a bit, but could protect you from the race condition in the long run.

- Test, test, test. Run your design through every conceivable scenario, trying to make it fail. This suggestion applies to everything you do in development, but it is particularly important when riding the edge with clocks. I have found some pretty obscure race conditions this way.
- Run as few clocks as possible. If you can perform all of your timings off of a single clock, it will pay dividends in power consumption.

2) Twist the Hardware.

Often, the microcontroller is among the smallest power consumer on the board before any optimization is performed. If this is the case, saving those precious microamperes is best accomplished in the hardware. This is one of the reasons so many embedded systems developers come from an engineering background, rather than computer science. Some suggestions:

- Make hardware switchable, where possible. The best way to minimize the power draw for a specific circuit is to turn it off when it is not in use. This is not always possible, and requires extra hardware and I/O, but is usually worth the trade-off. In most designs, it can be accomplished with a single transistor and I/O pin. If that gate and pin can extend battery life of your product by 10%, it certainly bears consideration.
- Play with the voltages. The '430 uses less current when run at 2.2 volts than at 3.6 volts. Along with device power, moving down to 2.2 VDC will also lower the output pin voltages, resulting in less current on pull-ups and other circuit devices. Remember, though, that the device won't function properly at maximum frequency and minimum supply voltage. Because of this, it sometimes makes sense to speed up the clocks, for a shorter period of time, rather than run them as slow as possible. It all boils down to understanding the circuit, including the microcontroller, and where the interactions lie.

3) Become a Systems Engineer.

This rule is less well-defined than the others. Often, the processor and hardware are both independently optimized for power, but there are trade-offs that can be made between the two that make the system more efficient. Changes that produce small increases in power consumption in the processor and surrounding circuitry can have more significant savings in other parts of the overall system. Simply knowing and meeting the firmware specs is not enough. In order to really optimize power, the diligent firmware developer understands the system as a whole, and how the processor fits. Ideally, optimization at this level occurs before circuit and software level fixes, but I never completely trust the systems engineers.

Batteries

If you are particularly interested in trying to squeeze out every possible bit of power consumption, it is likely that your design is operated off of a battery. Some helpful hints to remember about the battery in your design:

- You probably do not need to understand the chemistry of the battery, but you do should grasp its implications. Some battery chemistries are characterized by a gradual rolloff of voltage level as the battery power is consumed, while others remain rock solid at nominal voltage until there are only a few milliamp-hours of charge remaining, and then fall off the table. Maximum current, internal resistance, and efficiency will all change with aging, temperature, and temperature cycling.
- Battery lifetimes are usually specified in ampere-hours (i.e., how many hours the battery will last at 1 ampere). Keep in mind that this is not a fixed value for all batteries of this type. Rather, it is the mean of an admittedly narrow distribution of expected battery lifetimes.
- Batteries are almost always very stable power supplies. However, if your design is using replaceable or rechargeable batteries, your design will probably need a reset chip, or something similar, rather than an RC reset circuit. Remember, this device is highly sensitive to brownout conditions.

CHAPTER 13

A Sample Application

Now that the functions and features of the devices have been described, we will examine the development of a sample application. We will develop an 8-channel digital household lighting controller, capable of independently turning different lights and appliances in your home on and off independently, at any interval. We will use the MSP430F149, as there is a low-cost TI FET tool based on the device.

There are several initial design components we must consider. They are:

1) Event timing. The design will require a time counter, which increments each second. While this design could be performed with a full RTC, that would be design overkill, which I prefer to avoid. We will use a 32.768-kHz crystal oscillator to drive Timer A, which will be operated in count up mode. On overflow, the timer will generate an interrupt, at which point time-based event decisions will be made. The CPU will be run off of the DCO clock.

2) Interfacing to lighting and appliances. We will be using Port 4 to control devices, with each bit controlling an interface unit. An obvious problem lies in that of current sourcing. The MSP430F149 I/O pins can only source several milliamps of current, which is not enough for any switch or relay large enough to handle the power requirements of household devices. To handle this, each output pin will drive the base of a transistor, which will in turn drive an electro-

mechanical relay. This circuit is far from optimal, but it will serve its purpose for our design. Optimization of this type of circuit is a topic for another book.

3) Programming the time intervals. There needs to be a simple scheme to let the device know how long to leave devices on and off. The design will be for an interrupt-driven process, using the Port 1 interrupts to turn counters on and off. Port 2 will indicate when an individual channel is being programmed, and Port 3 will indicate whether it is timing an on-time interval, or an off-time interval. These LED ports are configured as active low, with the diode lighting up when a 0 is written to the pin.

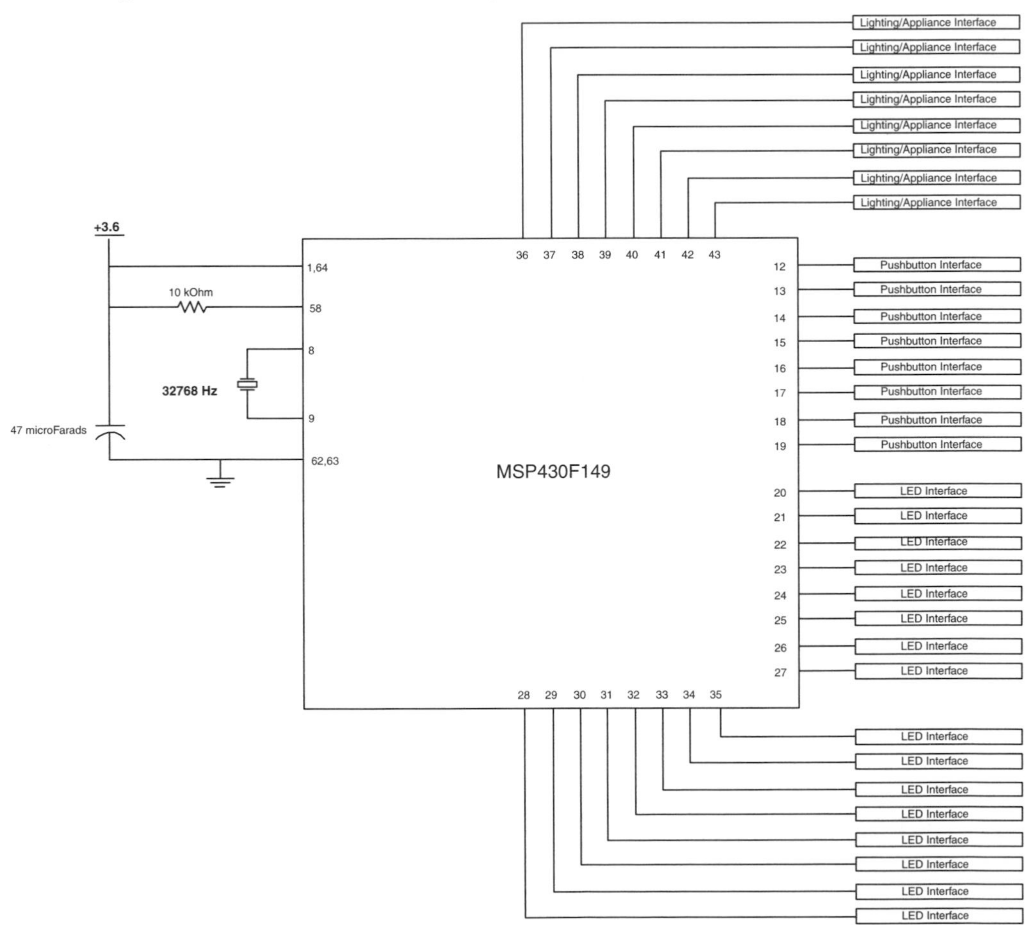

Figure 13.1: Lighting and Appliance Controller Schematic

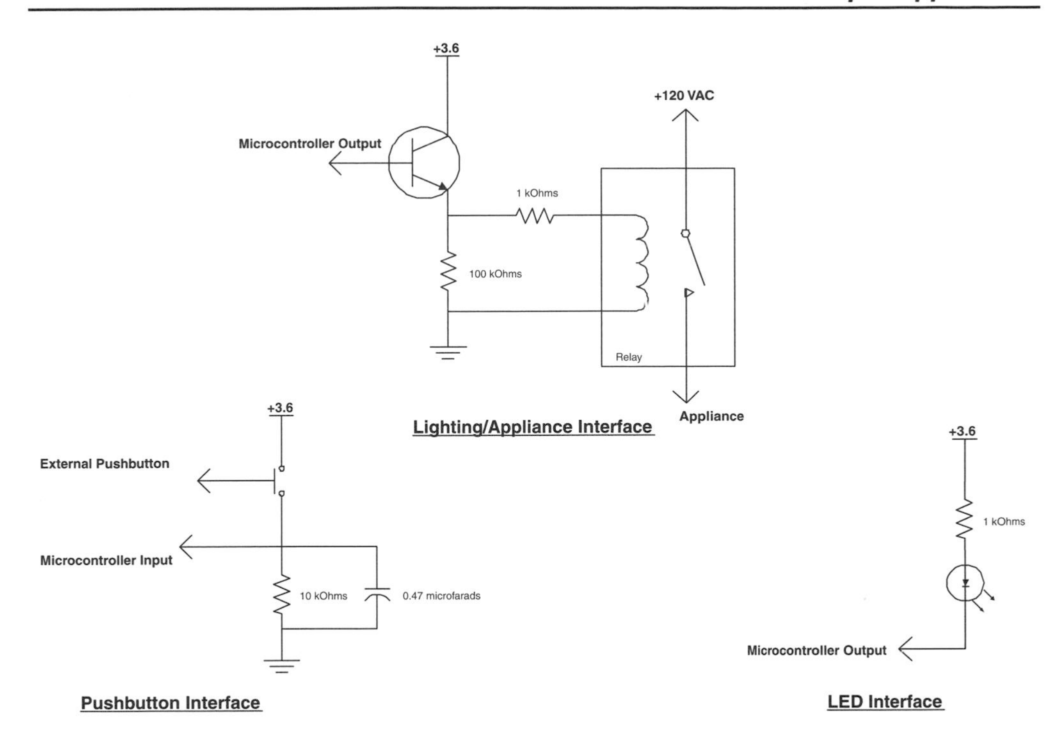

Figure 13.2: Lighting and Appliance Controller Interface Units

Hardware consists of the microcontroller (along with the associated crystal and reset circuit), eight interface units, eight pushbutton units, and sixteen LED indicator units, to reflect programming mode.

The code will perform some basic setup operations, and then enter an eternal main loop. All processing will be performed within the interrupt service routines. Typically, this is not the best way to handle things. However, we can handle the Port 1 interrupt with a single C instruction, and we will run the DCO at around 4 MHz, allowing for an awful lot of instruction cycles between 1 second interrupts from Timer A. Aside from these, we do not have other competing processes or interrupts to cause race conditions, allowing for plenty of timing margin.

Setup

Device initialization is performed by the function initializeDevice(). The initialization steps, in order, are:

1) Configure the watchdog timer. Due to the relative simplicity of this application, we will operate without a safety net, and turn the WDT off here.

2) Configure the clocks. The design will set the DCO to its maximum rate, from which the MCLK, and the CPU are operated. ACLK will be generated by the 32.768 kHz crystal, and will be the source for Timer A.

3) Configure the I/O ports. This includes interrupt enables for Ports 1 and 2, direction registers, output registers, and function registers. Since we are not using Port 5 for anything, we can set the function register elements, and bring ACLK and MCLK out, as a troubleshooting tool.

4) Set Timer A up. The timer will be clocked by ACLK. We will set Timer A into continuous up mode, and set the TAIE bit, so that the interrupt is generated when the timer overflows from 0xFFFF to 0x0000. These interrupts should happen every second, plus or minus any crystal oscillator error. In this configuration, we will not require use of any of the Capture/Compare modules. I typically set the peripherals up last, so that they are less likely to generate interrupts while still in the initialization routine. (An alternative is to set the Timer A up earlier, in stop mode, and start it later with a TACTL |= MC_2 command.) In this particular application, with the CPU clock running several orders of magnitude faster than the Timer A clock, it doesn't really matter. However, like most developers, I am a creature of habit, and respecting clocks and interrupts is a pretty good habit to have.

5) You will probably notice from the code below that I have initialized some registers to their default values (e.g., set P1DIR to 0x00 when it

should, on device reset, already be at 0x00). Again, this is one of those habits I have developed. As this is an instruction called at reset only, and we are nowhere near being constrained in terms of clock cycles or code space, this instruction is basically free. There is no reason not to make absolutely certain that registers are initialized to the correct values.

Code Listing 13.1 shows the function initializeDevice().

Listing 13.1: Device initialization function

```
void initializeDevice(void)
    {
    //Basic Clock Module Setup
    WDTCTL = 0x5A00 + WDTHOLD;     //sets WDT to hold
    DCOCTL =  0xE0;                //sets DCO to max frequency
    BCSCTL1 = XT2OFF + RSEL2 + RSEL1 + RSEL0;
                                   //no xt2 osc., max dco resistor
    BCSCTL2 = SELM_1;              //sets CPU to run off DCO

    //Initialize Ports
    P1DIR = 0x00;                  //Inputs on Port 1
    P1IES = 0xFF;                  //Interrupt on high-to-low
    P1IE = 0xFF;                   //Port 1 is interruptible
    P1SEL = 0x00;                  //No functions

    P2DIR = 0xFF;                  //Outputs on Port 2
    P2OUT = 0xFF;                  //Initialize to all 1.Active low output
    P2IE = 0x00;                   //Port 2 is not interruptible
    P2SEL = 0x00;                  //No functions

    P3DIR = 0xFF;                  //Outputs on Port 3
    P3OUT = 0xFF;                  //Initialize to all 1.Active low output
    P3SEL = 0x00;                  //No functions
```

```
        P4DIR = 0xFF;                       //Outputs on Port 4
        P4OUT = 0x00;                       //Initialize to all 0.Active high output
        P4SEL = 0x00;                       //No functions

        P5DIR = 0xFF;                       //Outputs on Port 5
        P5OUT = 0x00;                       //Initialize to all 0.
        P5SEL = 0xFF;                       //Functions for troubleshooting

        P6DIR = 0x00;                       //Inputs on Port 6
        P6SEL = 0x00;                       //No functions

        //Set Up Timer A
        TACTL = TASSEL_1 + MC_2 + TAIE;  //Sourced By ACLK, Up
                                          //mode, Interrupt En.

    return;
    }//End initializeDevice Function
```

Main Loop

Since all processing after setup is performed within the ISRs, we will use the proven and time-honored eternal loop operating system to run the device. Personally, I prefer to use the statement "while(1);" for the loop until the end of time function, but you have probably seen similar incarnations such as "for (;;);", or "jmp $" in assembly language. They are all good. Some applications use a very popular derivative of this that set flags within ISRs, and then repetitively perform checks of those flags with statements inside the eternal loop.

The eternal loop is popular for several reasons. First, it is predictable. Code seldom wanders off to an unplanned end of file with this type of structure. Second, and more importantly, this structure allows for the processor to enter sleep mode when not processing the ISRs, thereby saving power.

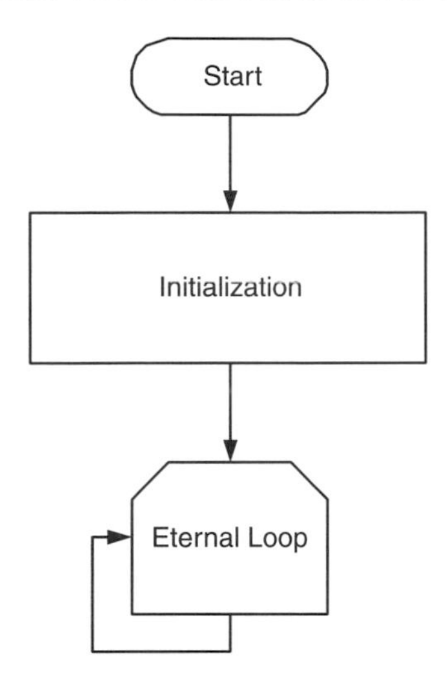

Figure 13.3: Main Program Loop

Interrupt Service Routines

There are two interrupt service routines to write. The first, which occurs every second on Timer A overflow, implements the flowchart in Figure 13.4. It automatically decrements each of eight individual seconds counters. If any of these counters runs out, it toggles the appropriate pin, and determines the new countdown value based on a bit flag, identified as a bit in the byte variable On_Time_Flag (with channel 0 having its flag at bit 0, channel 1 at bit 1, etc). The variables Timing_Underway_Flag and On_Time_Count_Flag are implemented identically. Since these variables are used in ISRs, they are, by necessity, global. (Software engineering types will rant about the use of global variables being evil, but in interrupt-driven microcontrollers, they are necessary and, in fact, proper.)

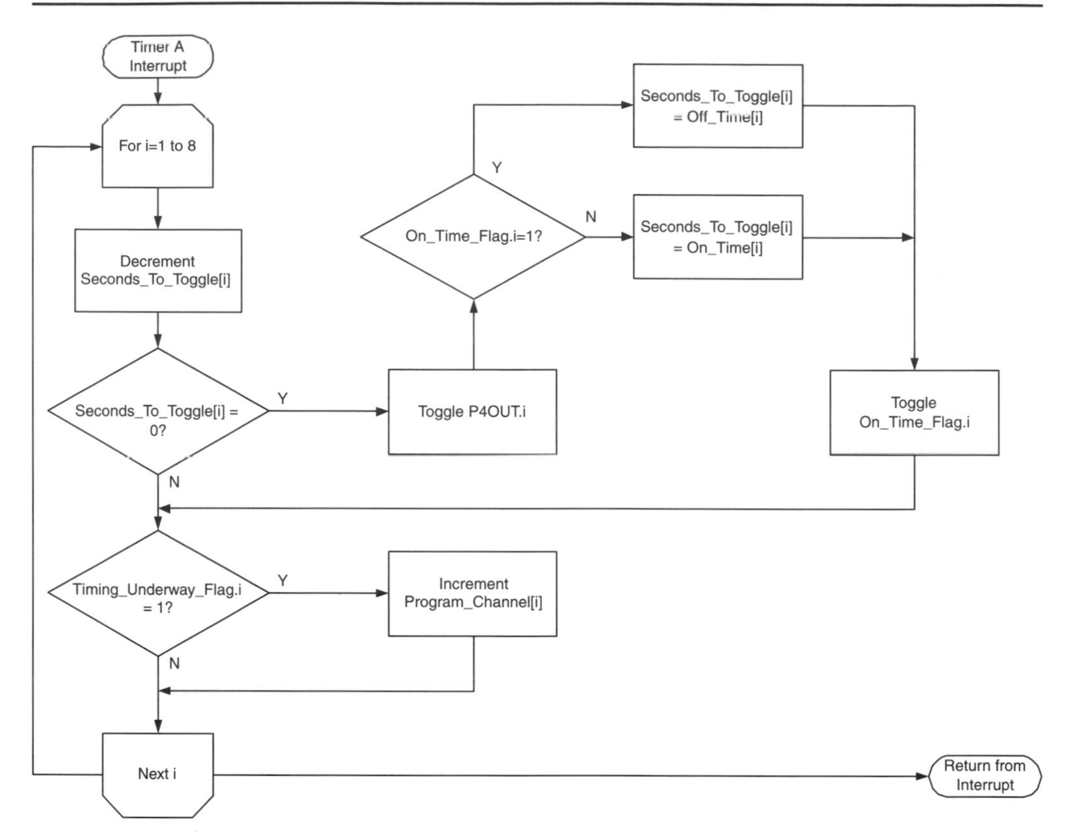

Figure 13.4. Timer A Interrupt Service Routine

Listing 13.2: Timer A Interrupt Service Routine

```
interrupt [0x0A] void TimerA_Interrupt (void)
    {
    unsigned char Bit_Mask = 0x01;      //Used to identify current bit in
                                        //flag fields
    unsigned int Loop_Index=0;          //Generic Loop Counter

    for (Loop_Index=0;Loop_Index<8;LoopIndex++)
        {
        Seconds_To_Toggle[Loop_Index]—;
        if (!(Seconds_To_Toggle[Loop_Index])  //Check for zero seconds
```

```
        {
        P4OUT ^= Bit_Mask;                   //Toggle Output Bit
        if (On_Time_Flag & Bit_Mask)         //On or Off??
            Seconds_To_Toggle[Loop_Index]=On_Time[Loop_Index];
        else
            Seconds_To_Toggle[Loop_Index]=Off_Time[Loop_Index];
        On_Time_Flag ^= Bit_Mask;            //Toggle On Time Flag
        }//End If
    Bit_Mask <<=1;
    if (Timing_Underway_Flag & Bit_Mask)  //If setting up channel
        Program_Channel[LoopIndex]++;     //Increment counter
    }//End For Loop
return;
}//End Timer A Interrupt Function
```

The second interrupt service routine is for programming the device, via Port 1 and the pushbutton interfaces described in Figure 13.2. To program the device, the user needs to define on and off times by pressing the specific channel button at the boundary times. For example, let's say the user wants to turn on a coffeepot, controlled by channel 7, at 8:00 every morning, and turn it off at 10:00 every morning. At 8:00 on the first day, the user pushes and releases button #7. The Programming Underway LED (on Port 2.7) and the On Time Programming LED (on Port 3.7) come on, and power is applied to the coffee maker, via the transistor/relay circuit on Port 4.7. At 10:00 that same day, the user pushes and releases button #7 again. The On Time Programming LED and the coffee maker both switch off, but the Programming Underway LED remains on. Finally, at 8:00 the following morning, the user presses and releases button #7 one final time. At this point, the Programming Underway LED switches off, the coffee maker switches on, and the channel is programmed and should operate correctly.

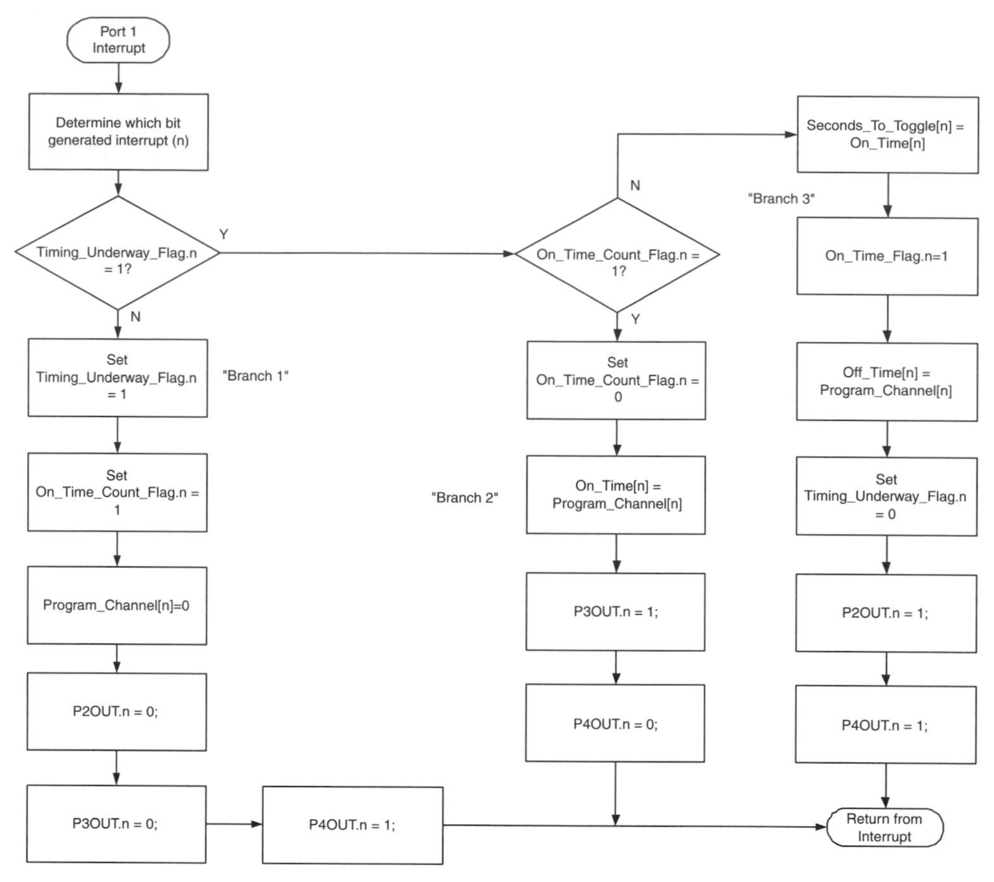

Figure 13.5: Port 1 Interrupt Service Routine

Listing 13.3: Port 1 Interrupt Service Routine

```
interrupt [0x08] void Port1_Interrupt (void)
   {
   unsigned char Bit_Mask = 0x00;      //Used to identify current bit in
                                       //flag fields
   unsigned int Loop_Index=0;          //Generic Loop Counter
   unsigned char Bit_Number=0x00;      //used as index into array for
                                       //setting values

   Bit_Mask=P1IFG;                     //Bit mask is identical to flag
                                       //register
```

```
//This for loop maps the bit number of the set bit in P1IFG to an
//integer value
for (Loop_Index=0;Loop_Index<8;Loop_Index++)
    {
    if ((P1IFG>>Loop_Index) & 0x01)  Bit_Number=Loop_Index;
    }//End For

if (Timing_Underway_Flag & Bit_Mask)
    {
    if (On_Time_Count_Flag & Bit_Mask)
                                //Branch 2 on the accompanying
                                //flowchart
        {
        On_Time_Count_Flag &= ~Bit_Mask;
                                //Clear On Time Flag
        On_Time[Bit_Number]=Program_Channel[Bit_Number];
        Program_Channel[Bit_Number]=0;
        P3OUT  |= Bit_Mask;
        P4OUT &= ~Bit_Mask;
        }//End inner if
    else                        //Branch 3 on the accompanying
                                //flowchart
        {
        Seconds_To_Toggle[Bit_Number] = On_Time[Bit_Number];
        On_Time_Flag  |= Bit_Mask;
        On_Time[Bit_Number]=Program_Channel[Bit_Number];
        Program_Channel[Bit_Number]=0;
        Timing_Underway_Flag &= ~Bit_Mask;
        P2OUT  |= Bit_Mask;
        P4OUT  |= Bit_Mask;
        }//End inner else
    }//End Outer If
```

```
else                              //Branch 1 on the accompanying flowchart
    {
    Timing_Underway_Flag |= Bit_Mask;
                                  //Set timing underway bit flag
    On_Time_Count_Flag |= Bit_Mask;
                                  //Set On Time count bit flag
    Program_Channel[Bit_Number]=0;
                                  //Initialize Counter
    P2OUT &= ~Bit_Mask;           //Clear appropriate bits on
    P3OUT &= ~Bit_Mask;           //  P2OUT and P3OUT
    P4OUT |= Bit_Mask;            //Turn on appliance switch
    }//End Else
return;
}//End Port 1  Interrupt Function
```

Putting It All Together

At this point, the bulk of the code is written. Once we make a couple of decisions about variable sizes, we can put it all together into our code set. We need to decide on the optimal length of variables, and declare them at the beginning of the code:

- Flags (Timing_Underway_Flag, On_Time_Count_Flag, and On_Time_Flag). Since we are implementing eight channels, these should obviously be byte elements. Since we are not performing computations based on these values, we will identify them each as type unsigned int.
- Second Counts (Seconds_To_Toggle, On_Time, Off_Time, and Program_Channel). This is not as straightforward. 16 bits is insufficient, as it only allows for counts up to about 18 hours. 32 bits is a bit of overkill, as it allows for periods of 136 years. Anything in the middle, however, would require us to define a special data type, which I typically prefer not to do, unless RAM is closely constrained. As this is not the case, we will use 32-bit, or unsigned long values, for these variables.

The final code set is in Listing 13.4.

Listing 13.4 Entire Code Set

```
/*************************************************************
    File:CodeSet.c
    Rev:1.0
    Date:1-1-03
    Author:CJN
*************************************************************/
 #include         "msp430x14x.h"
/*************************************************************
            Function Prototypes
*************************************************************/

void initializeDevice(void);



/*************************************************************
            Global Variables
*************************************************************/
unsigned char      Timing_Underway_Flag=0;
unsigned char      On_Time_Flag=0;
unsigned char      On_Time_Count_Flag=0;
unsigned long      Seconds_To_Count[7];
unsigned long      On_Time[7];
unsigned long      Off_Time[7];
unsigned long      Program_Channel[7];
```

```
/**************************************************************
        Main Program
**************************************************************

void main(void)
      {
      initializeDevice();
      while (1)
         {}//End Eterna-While
      }//End Main


/**************************************************************
        ISRs
**************************************************************/
interrupt [0x0A] void TimerA_Interrupt (void)
   {
   unsigned char Bit_Mask = 0x01;       //Used to identify current bit in
                                        //flag fields
   unsigned int Loop_Index=0;           //Generic Loop Counter

   for (Loop_Index=0;Loop_Index<8;LoopIndex++)
     {
     Seconds_To_Toggle[Loop_Index]—;
     if (!(Seconds_To_Toggle[Loop_Index])    //Check for zero seconds
         {
         P4OUT ^= Bit_Mask;                  //Toggle Output Bit
         if (On_Time_Flag & Bit_Mask)        //On or Off??
            Seconds_To_Toggle[Loop_Index]=On_Time[Loop_Index];
         else
            Seconds_To_Toggle[Loop_Index]=Off_Time[Loop_Index];
         On_Time_Flag ^= Bit_Mask;           //Toggle On Time Flag
         }//End If
```

```
        Bit_Mask <<=1;
        if (Timing_Underway_Flag & Bit_Mask) //If setting up channel
            Program_Channel[LoopIndex]++;    //Increment counter
        }//End For Loop
    return;
    }//End Timer A Interrupt Function

interrupt [0x08] void Port1_Interrupt (void)
    {
    unsigned char Bit_Mask = 0x00;       //Used to identify current bit in
                                         //flag fields
    unsigned int Loop_Index=0;           //Generic Loop Counter
    unsigned char Bit_Number=0x00;       //used as index into array for
                                         //setting values

    Bit_Mask=P1IFG;                      //Bit mask is identical to flag register

    //This for loop maps the bit number of the set bit in P1IFG to an integer value
    for (Loop_Index=0;Loop_Index<8;Loop_Index++)
        {
        if ((P1IFG>>Loop_Index) & 0x01) Bit_Number=Loop_Index;
        }//End For

    if (Timing_Underway_Flag & Bit_Mask)
        {
        if (On_Time_Count_Flag & Bit_Mask) //Branch 2 on the
                                           //accompanying flowchart
            {
            On_Time_Count_Flag &= ~Bit_Mask; //Clear On Time Flag
            On_Time[Bit_Number]=Program_Channel[Bit_Number];
            Program_Channel[Bit_Number]=0;
            P3OUT |= Bit_Mask;
            P4OUT &= ~Bit_Mask;
            }//End inner if
```

```
        else                                //Branch 3 on the accompanying flowchart
            {
            Seconds_To_Toggle[Bit_Number] = On_Time[Bit_Number];
            On_Time_Flag | = Bit_Mask;
            On_Time[Bit_Number]=Program_Channel[Bit_Number];
            Program_Channel[Bit_Number]=0;
            Timing_Underway_Flag &= ~Bit_Mask;
            P2OUT  | = Bit_Mask;
            P4OUT  | = Bit_Mask;
            }//End inner else
        }//End Outer If
    else                                    //Branch 1 on the accompanying flowchart
        {
        Timing_Underway_Flag | = Bit_Mask;  //Set timing underway bit
                                            //flag
        On_Time_Count_Flag | = Bit_Mask;    //Set On Time count bit
                                            //flag
        Program_Channel[Bit_Number]=0;      //Initialize Counter
        P2OUT &= ~Bit_Mask;                 //Clear appropriate bits on
        P3OUT &= ~Bit_Mask;                 //P2OUT and P3OUT
        P4OUT | = Bit_Mask;                 //Turn on appliance switch
        }//End Else
    return;
    }//End Port 1  Interrupt Function


/*************************************************************
            Functions
*************************************************************/


void initializeDevice(void)
    {
```

```
//Basic Clock Module Setup
WDTCTL = 0x5A00 + WDTHOLD;//sets WDT to hold
DCOCTL =  0xE0;                  //sets DCO to max frequency
BCSCTL1 = XT2OFF + RSEL2 + RSEL1 + RSEL0;
                                 //no xt2 osc., max dco resistor
BCSCTL2 = SELM_1;                //sets CPU to run off DCO

//Initialize Ports
P1DIR = 0x00;                    //Inputs on Port 1
P1IES = 0xFF;                    //Interrupt on high-to-low
P1IE = 0xFF;                     //Port 1 is interruptible
P1SEL = 0x00;                    //No functions

P2DIR = 0xFF;                    //Outputs on Port 2
P2OUT = 0xFF;                    //Initialize to all 1. Active low output
P2IE = 0x00;                     //Port 2 is not interruptible
P2SEL = 0x00;                    //No functions

P3DIR = 0xFF;                    //Outputs on Port 3
P3OUT = 0xFF;                    //Initialize to all 1. Active low
                                 //output
P3SEL = 0x00;                    //No functions

P4DIR = 0xFF;                    //Outputs on Port 4
P4OUT = 0x00;                    //Initialize to all 0. Active high
                                 //output
P4SEL = 0x00;                    //No functions

P5DIR = 0xFF;                    //Outputs on Port 5
P5OUT = 0x00;                    //Initialize to all 0.
P5SEL = 0xFF;                    //Functions for troubleshooting
```

```
P6DIR = 0x00;                          //Inputs on Port 6
P6SEL = 0x00;                          //No functions

//Set Up Timer A
TACTL = TASSEL_1 + MC_2 + TAIE;
                                       //Sourced By ACLK, Up mode,
                                       //Interrupt En.

return;
}//End initializeDevice Function
```

What is wrong with this design?

This is a useful question to ask about any design, after the first cut. There are several sources of potential misbehavior on the part of this design:

- *Multiple Port 1 interrupts*. If there is more than 1 bit set in P1IFG when the ISR is kicked off, the Bit_Mask variable will have several bits set, while the Bit_Number variable will only reflect the most significant set bit. While this will cause some strange (and incorrect) behavior, it should be uncommon enough that it is probably sufficient that we are aware of it, but can live with it.

- *Supply and reset stability*. For many power supplies, particularly a cheap wall-transformer type DC supply, the RC circuit I have put in for the reset circuit is horribly insufficient. Anytime there is a lengthy enough power loss or glitch on the reset line, the device will need to be reprogrammed from scratch. Make the power and reset signal as stable as possible.

- *Crystal Error*. Remember, timed events are never more accurate than the crystal from which the time base is derived. If your crystal has an error of 150 ppm (parts per million, a common unit of measure for crystals), an event time will drift by about 90 minutes per year.

Some Functional Additions

This design, along with the rich feature set of the '430 devices, allows for some obvious additions. Rather than switch outputs on and off based on time, they can be switched based on temperature or external signal levels, using the A/D convertor and the comparator circuits. There are some considerations when implementing these:

- Pin use. Currently, Port 2 is used for indicators. However, the comparator function pins are co-located on this port. If a comparator input is used, this indicator will need to be eliminated or co-located.
- Interrupt use. Since the functionality is already resident within an ISR, using interrupt driven comparator or A/D conversion won't work without forcing re-entrant interrupt, a subject discussed in Chapter 3. We will code to avoid this.
- Debounce. If either condition is near the defined threshold, checking every second could easily cause an oscillation. There are two easy alternatives to this. The frequency of condition checking can be set low enough (such as 30 minutes) that any oscillation is acceptable. The second alternative is to implement hysteresis, by changing the threshold after a state change. Neither choice is without significant disadvantage (latency vs. shifting parameters). We will implement the first choice.

There is a single variable added to the code, Seconds_To_Condition_Check, which is implemented as an unsigned int, and initialized to 1800 (which is equivalent to 30 minutes). The implementation uses the peripheral code developed in Chapter 6 for temperature sensing and comparator decisions. The temperature threshold is hard-coded in, but could be implemented as a dynamic variable. As the ISR is written to use channels 7 and 8 for the temperature and comparator channels, the time checking loop that follows has been limited to the first six channels.

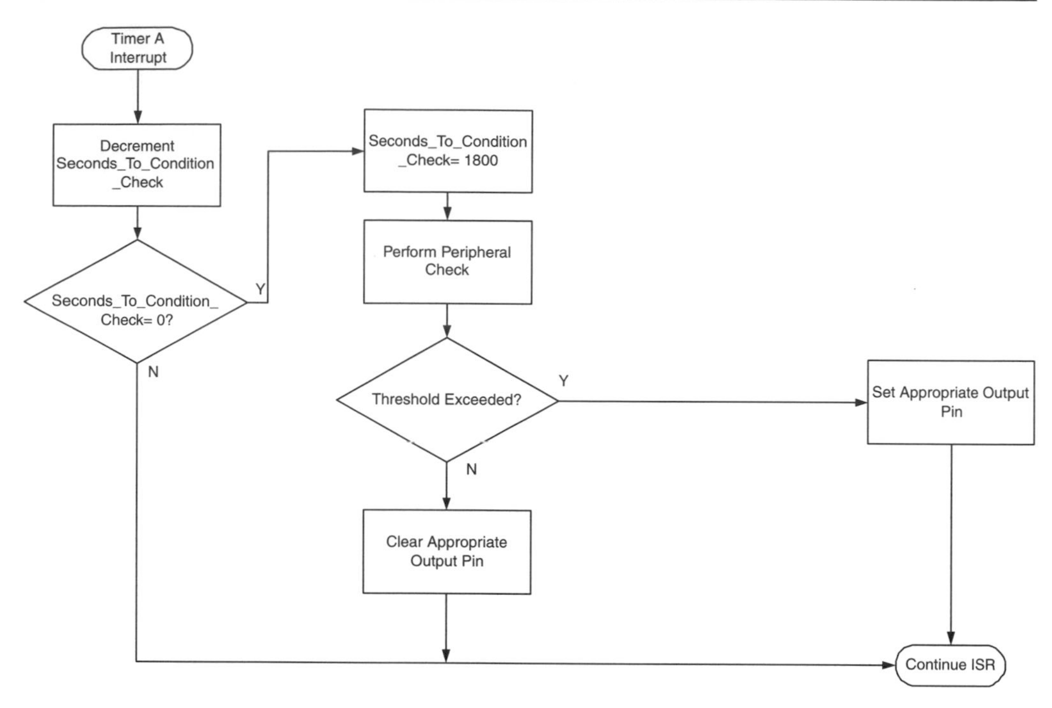

Figure 13.6: Additions to Timer Interrupt

Listing 13.5: Timer Interrupt With Comparator and Temperature Modifications

```
interrupt [0x0A] void TimerA_Interrupt (void)
    {
    unsigned char Bit_Mask = 0x01;      //Used to identify current bit in
                        //flag fields
    unsigned int Loop_Index=0;          //Generic Loop Counter
    signed long Temperature=25;

    Seconds_To_Condition_Check—;
    if (!(Seconds_To_Condition_Check))
        {
        Seconds_To_Condition_Check = 1800;
```

```
/*****************Comparator Decision***********************/
        P2SEL |= 0x10;                      //Set up I/O direction register
        CACTL1 = CAREF_21 + CAON;
                                            //Turn on Comparator, internal
                                            //Vcc/2 reference
        CACTL2 = CAF + P2CA1;               //Use external signal for –
                                            //comparator input
        CAPD = 0xFD;                        //Enable CAPD1 input buffer
        for (Loop_Index =0; Loop_Index <15; Loop_Index ++) {}
                                            //Loop delay allows filter output
                                            //to stabilize
                                            //the length will depend on clock
                                            //speed
        if (CACTL2 & CAOUT)
            {
                P4OUT |= 0x80;              //Set Pin High
            }
        else
            {
                P4OUT &= ~0x80;             //Clear Pin
        }

/*****************Temperature Decision*********************/

        ADC12CTL0 = SHT0_6 + SHT1_6 + REFON + ADC12ON;
                //Initialize control register 0. Sets up conversion clock,
                //internal reference
        ADC12CTL1 = SHP;
                //Conversion sample timer signal is sourced from the
                //sampling timer
        ADC12MCTL0 = INCH_10 + SREF_1;
                //Use internal reference with respect to ground, Temp on
                //input channel 10.
```

```
        ADC12CTL0 | = ADC12SC + ENC;    //Enable and start
                                        //conversion
        while (ADC12CTL1 & 0x01);       // Hang in loop until
                                        //conversion completes.
       Temperature = ADC12MEM0 & 0x00000FFF;
       Temperature  *= 845;          //These steps convert the ADC
                                     //reading to degrees Celsius

       Temperature  >> = 13;
       Temperature  -= 278;
       if (Temperature>35)
           {
               P4OUT | = 0x40;       //Set Pin High
           }
       else
           {
          P4OUT &= ~0x40;            //Clear Pin
       }


       }//End Condition Check  if



   for (Loop_Index=0;Loop_Index<6;LoopIndex++)
       {
       Seconds_To_Toggle[Loop_Index]—;
       if (!(Seconds_To_Toggle[Loop_Index])  //Check for zero
                                             //seconds
           {
           P4OUT ^= Bit_Mask;                //Toggle Output Bit
           if (On_Time_Flag & Bit_Mask)      //On or Off??
           Seconds_To_Toggle[Loop_Index]=On_Time[Loop_Index];
```

```
    else
        Seconds_To_Toggle[Loop_Index]=Off_Time[Loop_Index];
    On_Time_Flag ^= Bit_Mask;              //Toggle On Time Flag
    }//End If
Bit_Mask <<=1;
if (Timing_Underway_Flag & Bit_Mask)      //If setting up channel
    Program_Channel[LoopIndex]++;          //Increment counter
    }//End For Loop
return;
}//End Timer A Interrupt Function
```

Firmware Testing

Among the most misunderstood and misused steps in the product development process is that of firmware testing. All too often, the testing of the firmware is assumed to be adequately performed by the unit testing process. A well-tested product checks the firmware itself, along with the interaction between firmware and hardware. Some specific examples:

- Code Coverage. This is the most important and often overlooked firmware test concept. The testing process needs to exercise ALL of the firmware, in various operational conditions. A surprising number of products have reached the marketplace, only to be found to have a firmware bug that presented itself in rare cases. Face it, if the bug were obvious, it never would have made it off the bench. Check all corners of your code, in every conceivable combination. Use automated tools to make life easier, if necessary.
- Temperature testing. Traditionally, this is considered to be a hardware test. Consider, however, the multiple clock sources that the MSP430 offers. They all vary in accuracy over temperature. However, the DCO varies by several orders of magnitude more than most crystals over the allowable temperature range. This can create cases where timings are solid at one temperature, but cause code crash at other temperature ranges. Run it over the entire specified range for your design.
- Reset/Supply/Brownout. Again, this is technically part of hardware test, but be aware that the MSP430 is VERY sensitive to brownouts, supply glitches, or any flakiness on the reset line. Use a decent surge tester to beat the crap out of your design, so that you will understand the failure modes.

APPENDIX A

Other Sources

There are many other good sources of information on these devices, and embedded systems development in general. Some of the sources that I have found to be most useful:

Texas Instruments

The website at *www.TI.com* is your most valuable source of information. Some of the most useful documents there are:

- SLAU049 MSP430x1xx Family Users Guide
- SLAU012 MSP430x3xx Family Users Guide
- SLAU056 MSP430x4xx Family Users Guide
- SLAA140 Application Note "Mixing C and Assembler with the MSP430"
- SLAA149 Application Note "Programming a Flash-Based MSP430 Using the JTAG Interface'
- SLAA089 Application Note "Features of the Bootstrap Loader"

This list is by no means inclusive; it merely represents the documentation I have found to be most useful. Take the time to review the contents of the website thoroughly.

Other Embedded Resources

- *Embedded Systems Programming* is a good magazine for picking up tips and ideas. It is a free monthly magazine. Its website is *www.embedded.com*.
- The newsgroup *comp.arch.embedded* has proven useful for picking up suggestions. The biggest advantage to newsgroups is that anyone can post anything, and the biggest disadvantage is that anyone can post anything. However, this group is one of the better ones that I read regularly, with most (but not all) posters providing honest and useful information.
- If possible, I strongly recommend attending one of the annual Embedded Systems Conferences. They are offered every spring in San Francisco, every summer in Chicago, and every fall in Boston. They are a good way to keep abreast of the recent changes in industry.

APPENDIX B

TI FET Tool

As discussed throughout the book, Texas Instruments offers a low-cost flash emulation tool, for use in prototyping and development. The kit includes the necessary hardware and software for developing and running code from a PC, including the popular IAR assembler and compiler. It is designed so that you can be developing within a few minutes of opening the box.

Kit Contents

The development kit contains:

- Two MSP430 microcontrollers.
- One printed-circuit board, with a clamshell type adapter socket for the microcontroller, external power and ground connections, a JTAG connector, several jumper connectors, pads for component growth, and an LED, for use in a demo program.
- A JTAG interface unit, which interfaces the signals from the PC serial port to the JTAG connection on the circuit board.
- Sets of pins and sockets. You will need to decide which you prefer to use for contacts, and solder them onto the circuit board, for external connections to the device. I always use the sockets, because if I change my mind later, the pin sets will sit in the sockets, giving me pins to use.
- The cables necessary to connect all this stuff together.

- A CD-ROM, which contains Kickstart, the development environment, and all of the FET documentation. Kickstart is a slightly modified version of IAR Systems' Embedded Workbench, which contains an assembler and compiler, along with C-SPY, which is a ROM monitor tool, that performs many standard emulation functions via the JTAG port.

Setting Up

Setting up is simple and straightforward, and is adequately described in the included documentation. I have, over several installations, discovered several quirks in the installation process:

- If you have had a previous installation of Kickstart or Embedded Workbench on the target PC, you need to eliminate all traces of it before the new install, including manually deleting the old IAR Systems folders. For some reason, a simple uninstall doesn't erase those, and the new installation doesn't seem to like the fact that they exist.
- The software does not always recognize the COM port the first time. When installing the device on COM1, I sometimes have to switch the software to COM2, and then back to COM1 before it finds the hardware. I do not know if this is a limitation of the software or the operating system, but it has always cleared up for me after one back-and-forth switch.
- Check the printed circuit configuration. It is initially set up for the device to be powered from the JTAG interface, with no external crystals, and a connection to the on-board LED. The LED is there for a demonstration program (which makes the LED blink) and needs to be disconnected before attempting anything useful with the device. The LED is connected via a jumper, and you should remove that.

Using Kickstart and the FET

As with the setup, usage of these tools is pretty well described by the included documentation. There are, however, a few tidbits that I have discovered, that might prove useful.

- The hardware will sometimes, when attempting to download new code, hang for no apparent reason. The software will report, via an error message in C-SPY, that it cannot find or recognize the hardware. It could be a brownout in the microcontroller (something that they are very sensitive to), or a latching error in the interface. I have found, however, that completely detaching the interface box, resetting the controller, and attaching everything back together tends to fix the problem.
- Included in Kickstart is C-SPY, the ROM monitor tool. It allows the FET tool to do some basic emulation functions. It is not terribly powerful compared to a full emulator, but will allow you to step through code and set a couple of simple breakpoints. (The full-cost version of EW/C-SPY offers the same feature level.)
- There is a setting that allows you to disconnect the JTAG while running the software. It is important to remember that, if you are performing any detailed timing, the JTAG can potentially screw that up. Selecting the "disconnect JTAG on go" feature will alleviate this problem, but you cannot step through the code with the JTAG disconnected.
- One function that is nice to have is the ability to print out the contents of memory, for later analysis. C-SPY will let you examine memory and CPU contents on the screen, but the copy and print functions are not implemented. Personally, I prefer to have printouts of this sort of thing, so that I can mark them up. The ability to review several thousand bytes of hex code on the screen is a dubious feature, at best. Fortunately, there is a way around this. In C-SPY, under the FET Options menu tab, under Advanced, select "Memory Dump". This feature allows you to select any portion of the device memory, and save it, with address information and the CPU registers, to a file. I always change the file extension from .dmp to .txt, so that any text editor can read (and print) the file.

APPENDIX C

Useful Acronyms

As engineers, we love our acronyms. Here is a short list of common acronyms from this book:

ADC	Analog-to-Digital Conversion or Converter
ALU	Arithmetic Logic Unit
BCD	Binary Coded Decimal
CG1/CG2	Constant Generators 1 or 2 are CPU registers R2 and R3, respectively.
CISC	Complex Instruction Set Computing
CMOS	Complementary Metal Oxide Semiconductor
CPU	Central Processing Unit
CRC	Cyclic Redundancy Check
DCO	Digitally Controlled Oscillator
EEPROM	Electrically Erasable Programmable Read Only Memory
FET	Flash Emulation Tool
FLL	Frequency Locked Loop
ISR	Interrupt Service Routine

LCD	Liquid Crystal Display
MAC	Multiply-and-Accumulate
MIPS	Million Instructions per Second
NMI	Non-Maskable Interrupt
OTP	One Time Programmable
PC	Program Counter
POR	Power On Reset
PUC	Power Up Clear
PWM	Pulse Width Modulator
RAM	Random Access Memory
RISC	Reduced Instruction Set Computing
ROM	Read Only Memory
RST	Reset
RTC	Real Time Clock
SFR	Special Function Registers. These include the PC (Program Counter), SP (Stack Pointer), and the SR (Status Register).
SP	Stack Pointer
SR	Status Register
TLA	Three Letter Acronym
TTL	Transistor-Transistor Logic
UART	Universal Asynchronous Receive/Transmit
USART	Universal Synchronous/Asynchronous Receive/Transmit
WDT	Watchdog Timer

APPENDIX D

A Sample Datasheet[1]

MSP430x13x, MSP430x14x
MIXED SIGNAL MICROCONTROLLER

- Low Supply-Voltage Range, 1.8 V . . . 3.6 V
- Ultralow-Power Consumption:
  - Active Mode: 280 μA at 1 MHz, 2.2V
  - Standby Mode: 1.6 μA
  - Off Mode (RAM Retention): 0.1 μA
- Five Power-Saving Modes
- Wake-Up From Standby Mode in 6 μs
- 16-Bit RISC Architecture, 125-ns Instruction Cycle Time
- 12-Bit A/D Converter With Internal Reference, Sample-and-Hold and Autoscan Feature
- 16-Bit Timer_B With Seven Capture/Compare-With-Shadow Registers
- 16-Bit Timer_A With Three Capture/Compare Registers
- On-Chip Comparator
- Serial Onboard Programming, No External Programming Voltage Needed Programmable Code Protection by Security Fuse
- Serial Communication Interface (USART), Functions as Asynchronous UART or Synchronous SPI Interface
  - Two USARTs (USART0, USART1) — MSP430x14x Devices
  - One USART (USART0) — MSP430x13x Devices
- Family Members Include:
  - MSP430F133: 8KB+256B Flash Memory, 256B RAM
  - MSP430F135: 16KB+256B Flash Memory, 512B RAM
  - MSP430F147: 32KB+256B Flash Memory, 1KB RAM
  - MSP430F148: 48KB+256B Flash Memory, 2KB RAM
  - MSP430F149: 60KB+256B Flash Memory, 2KB RAM
- Available in 64-Pin Quad Flat Pack (QFP)
- For Complete Module Descriptions, See the *MSP430x1xx Family User's Guide*, Literature Number SLAU049

description

The Texas Instruments MSP430 family of ultralow-power microcontrollers consist of several devices featuring different sets of peripherals targeted for various applications. The architecture, combined with five low power modes is optimized to achieve extended battery life in portable measurement applications. The device features a powerful 16-bit RISC CPU, 16-bit registers, and constant generators that attribute to maximum code efficiency. The digitally controlled oscillator (DCO) allows wake-up from low-power modes to active mode in less than 6 μs.

The MSP430x13x and the MSP430x14x series are microcontroller configurations with two built-in 16-bit timers, a fast 12-bit A/D converter, one or two universal serial synchronous/asynchronous communication interfaces (USART), and 48 I/O pins.

Typical applications include sensor systems that capture analog signals, convert them to digital values, and process and transmit the data to a host system. The timers make the configurations ideal for industrial control applications such as ripple counters, digital motor control, EE-meters, hand-held meters, etc. The hardware multiplier enhances the performance and offers a broad code and hardware-compatible family solution.

[1] Included with permission of Texas Instruments. For the most recent updates, visit www.TI.com.

MSP430x13x, MSP430x14x
MIXED SIGNAL MICROCONTROLLER

AVAILABLE OPTIONS

T_A	PACKAGED DEVICES
	PLASTIC 64-PIN QFP (PM)
–40°C to 85°C	MSP430F133IPM MSP430F135IPM MSP430F147IPM MSP430F148IPM MSP430F149IPM

pin designation, MSP430F133, MSP430F135

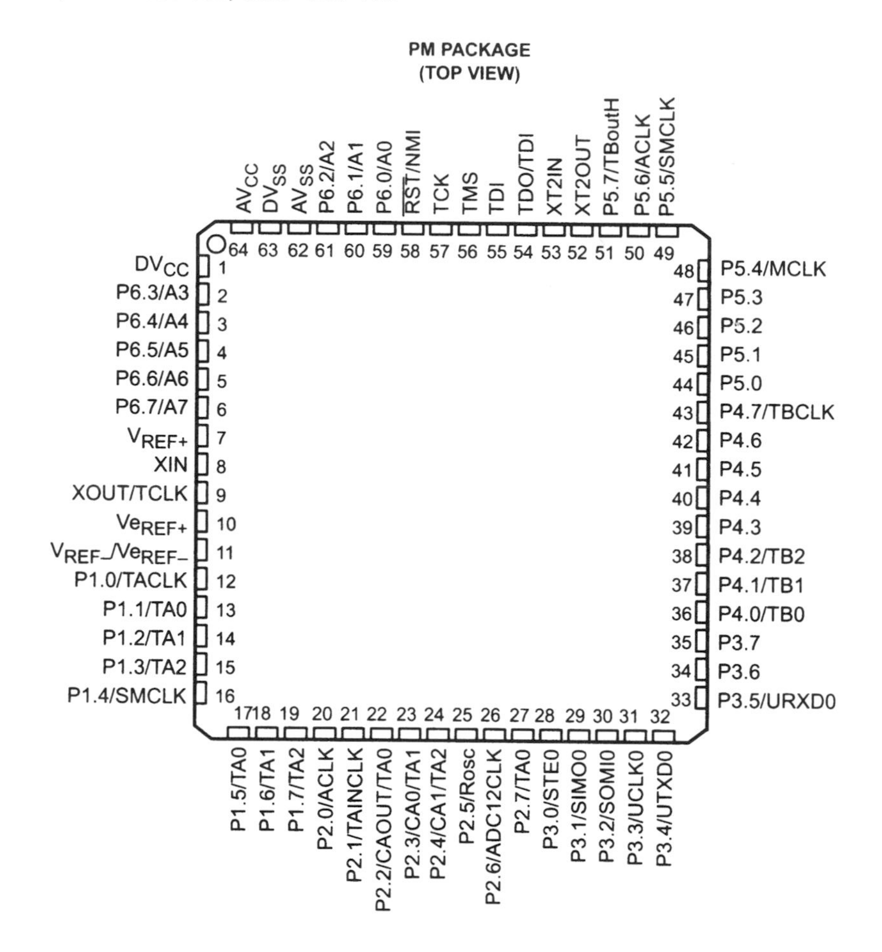

MSP430x13x, MSP430x14x
MIXED SIGNAL MICROCONTROLLER

pin designation, MSP430F147, MSP430F148, MSP430F149

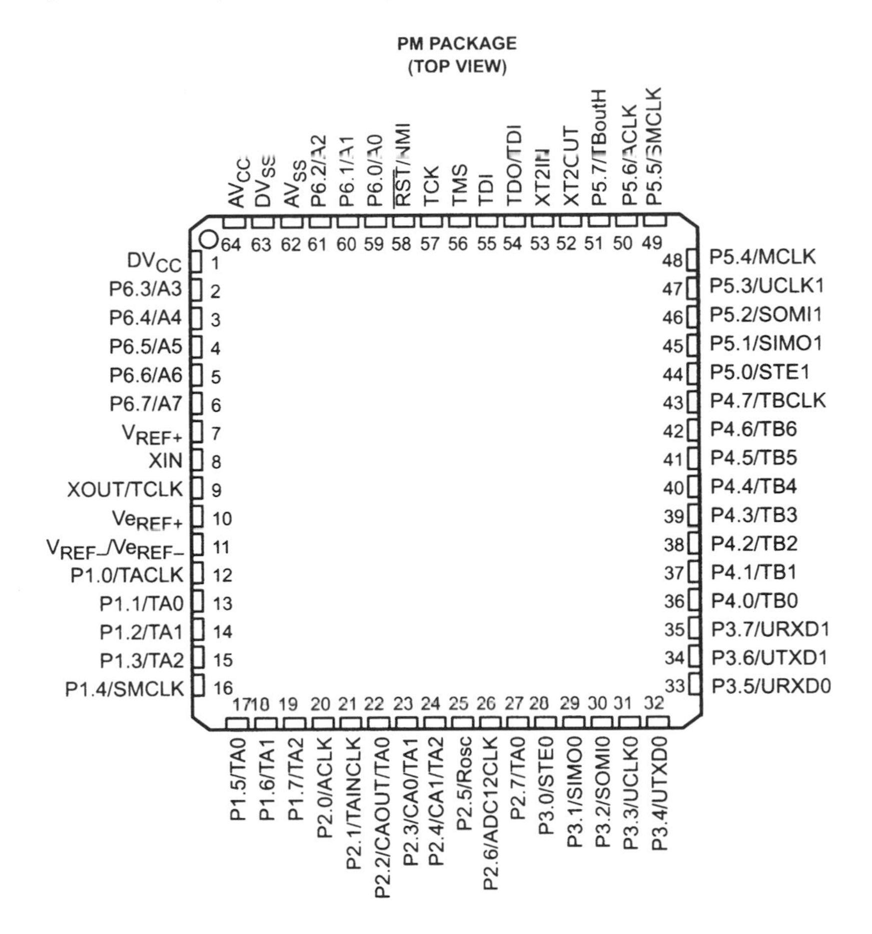

MSP430x13x, MSP430x14x
MIXED SIGNAL MICROCONTROLLER

functional block diagrams

MSP430x13x

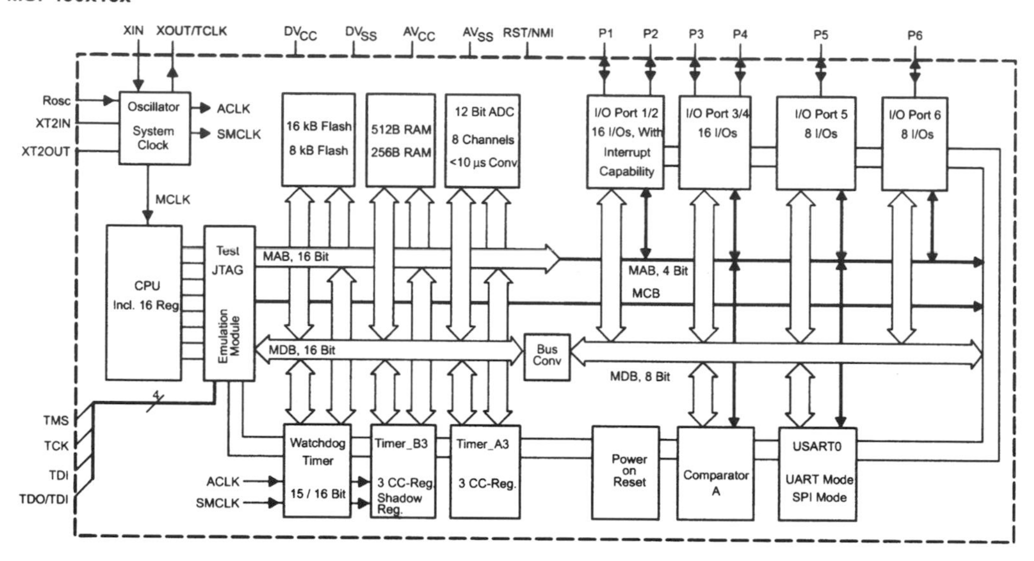

MSP430x14x

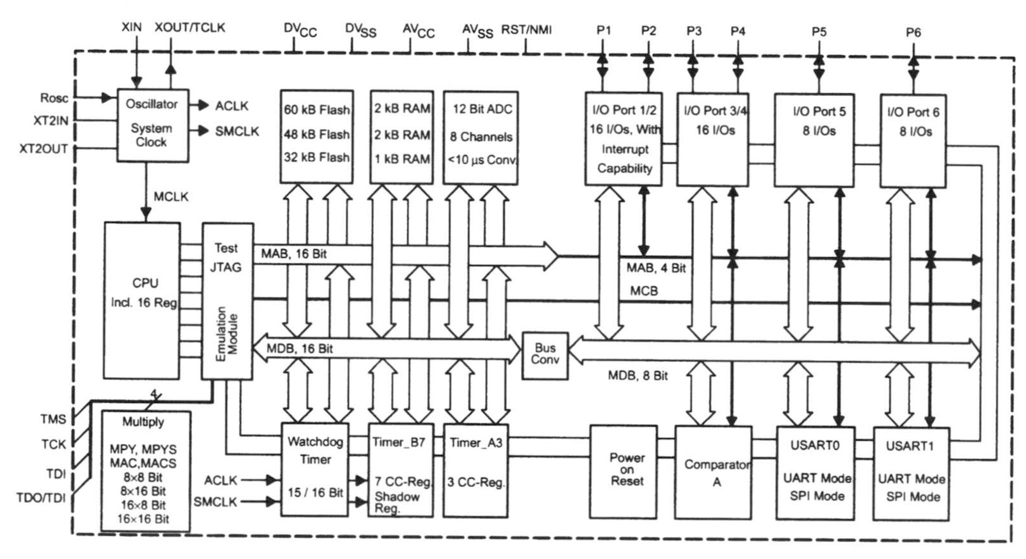

MSP430x13x, MSP430x14x
MIXED SIGNAL MICROCONTROLLER

Terminal Functions

TERMINAL NAME	NO.	I/O	DESCRIPTION
AV_{CC}	64		Analog supply voltage, positive terminal. Supplies only the analog portion of the analog-to-digital converter.
AV_{SS}	62		Analog supply voltage, negative terminal. Supplies only the analog portion of the analog-to-digital converter.
DV_{CC}	1		Digital supply voltage, positive terminal. Supplies all digital parts.
DV_{SS}	63		Digital supply voltage, negative terminal. Supplies all digital parts.
P1.0/TACLK	12	I/O	General digital I/O pin/Timer_A, clock signal TACLK input
P1.1/TA0	13	I/O	General digital I/O pin/Timer_A, capture: CCI0A input, compare: Out0 output
P1.2/TA1	14	I/O	General digital I/O pin/Timer_A, capture: CCI1A input, compare: Out1 output
P1.3/TA2	15	I/O	General digital I/O pin/Timer_A, capture: CCI2A input, compare: Out2 output
P1.4/SMCLK	16	I/O	General digital I/O pin/SMCLK signal output
P1.5/TA0	17	I/O	General digital I/O pin/Timer_A, compare: Out0 output
P1.6/TA1	18	I/O	General digital I/O pin/Timer_A, compare: Out1 output
P1.7/TA2	19	I/O	General digital I/O pin/Timer_A, compare: Out2 output/
P2.0/ACLK	20	I/O	General digital I/O pin/ACLK output
P2.1/TAINCLK	21	I/O	General digital I/O pin/Timer_A, clock signal at INCLK
P2.2/CAOUT/TA0	22	I/O	General digital I/O pin/Timer_A, capture: CCI0B input/Comparator_A output
P2.3/CA0/TA1	23	I/O	General digital I/O pin/Timer_A, compare: Out1 output/Comparator_A input
P2.4/CA1/TA2	24	I/O	General digital I/O pin/Timer_A, compare: Out2 output/Comparator_A input
P2.5/Rosc	25	I/O	General-purpose digital I/O pin, input for external resistor defining the DCO nominal frequency
P2.6/ADC12CLK	26	I/O	General digital I/O pin, conversion clock – 12-bit ADC
P2.7/TA0	27	I/O	General digital I/O pin/Timer_A, compare: Out0 output
P3.0/STE0	28	I/O	General digital I/O, slave transmit enable – USART0/SPI mode
P3.1/SIMO0	29	I/O	General digital I/O, slave in/master out of USART0/SPI mode
P3.2/SOMI0	30	I/O	General digital I/O, slave out/master in of USART0/SPI mode
P3.3/UCLK0	31	I/O	General digital I/O, external clock input – USART0/UART or SPI mode, clock output – USART0/SPI mode
P3.4/UTXD0	32	I/O	General digital I/O, transmit data out – USART0/UART mode
P3.5/URXD0	33	I/O	General digital I/O, receive data in – USART0/UART mode
P3.6/UTXD1†	34	I/O	General digital I/O, transmit data out – USART1/UART mode
P3.7/URXD1†	35	I/O	General digital I/O, receive data in – USART1/UART mode
P4.0/TB0	36	I/O	General-purpose digital I/O, capture I/P or PWM output port – Timer_B7 CCR0
P4.1/TB1	37	I/O	General-purpose digital I/O, capture I/P or PWM output port – Timer_B7 CCR1
P4.2/TB2	38	I/O	General-purpose digital I/O, capture I/P or PWM output port – Timer_B7 CCR2
P4.3/TB3†	39	I/O	General-purpose digital I/O, capture I/P or PWM output port – Timer_B7 CCR3
P4.4/TB4†	40	I/O	General-purpose digital I/O, capture I/P or PWM output port – Timer_B7 CCR4
P4.5/TB5†	41	I/O	General-purpose digital I/O, capture I/P or PWM output port – Timer_B7 CCR5
P4.6/TB6†	42	I/O	General-purpose digital I/O, capture I/P or PWM output port – Timer_B7 CCR6
P4.7/TBCLK	43	I/O	General-purpose digital I/O, input clock TBCLK – Timer_B7
P5.0/STE1†	44	I/O	General-purpose digital I/O, slave transmit enable – USART1/SPI mode
P5.1/SIMO1†	45	I/O	General-purpose digital I/O slave in/master out of USART1/SPI mode
P5.2/SOMI1†	46	I/O	General-purpose digital I/O, slave out/master in of USART1/SPI mode
P5.3/UCLK1†	47	I/O	General-purpose digital I/O, external clock input – USART1/UART or SPI mode, clock output – USART1/SPI mode
P5.4/MCLK	48	I/O	General-purpose digital I/O, main system clock MCLK output
P5.5/SMCLK	49	I/O	General-purpose digital I/O, submain system clock SMCLK output

† 14x devices only

MSP430x13x, MSP430x14x
MIXED SIGNAL MICROCONTROLLER

Terminal Functions (Continued)

TERMINAL NAME	NO.	I/O	DESCRIPTION
P5.6/ACLK	50	I/O	General-purpose digital I/O, auxiliary clock ACLK output
P5.7/TboutH	51	I/O	General-purpose digital I/O, switch all PWM digital output ports to high impedance – Timer_B7 TB0 to TB6
P6.0/A0	59	I/O	General digital I/O, analog input a0 – 12-bit ADC
P6.1/A1	60	I/O	General digital I/O, analog input a1 – 12-bit ADC
P6.2/A2	61	I/O	General digital I/O, analog input a2 – 12-bit ADC
P6.3/A3	2	I/O	General digital I/O, analog input a3 – 12-bit ADC
P6.4/A4	3	I/O	General digital I/O, analog input a4 – 12-bit ADC
P6.5/A5	4	I/O	General digital I/O, analog input a5 – 12-bit ADC
P6.6/A6	5	I/O	General digital I/O, analog input a6 – 12-bit ADC
P6.7/A7	6	I/O	General digital I/O, analog input a7 – 12-bit ADC
$\overline{RST}$/NMI	58	I	Reset input, nonmaskable interrupt input port, or bootstrap loader start (in Flash devices).
TCK	57	I	Test clock. TCK is the clock input port for device programming test and bootstrap loader start (in Flash devices).
TDI	55	I	Test data input. TDI is used as a data input port. The device protection fuse is connected to TDI.
TDO/TDI	54	I/O	Test data output port. TDO/TDI data output or programming data input terminal
TMS	56	I	Test mode select. TMS is used as an input port for device programming and test.
V_{eREF+}	10	I/P	Input for an external reference voltage to the ADC
V_{REF+}	7	O	Output of positive terminal of the reference voltage in the ADC
V_{REF-}/V_{eREF-}	11	O	Negative terminal for the ADC's reference voltage for both sources, the internal reference voltage, or an external applied reference voltage
XIN	8	I	Input port for crystal oscillator XT1. Standard or watch crystals can be connected.
XOUT/TCLK	9	I/O	Output terminal of crystal oscillator XT1 or test clock input
XT2IN	53	I	Input port for crystal oscillator XT2. Only standard crystals can be connected.
XT2OUT	52	O	Output terminal of crystal oscillator XT2

MSP430x13x, MSP430x14x
MIXED SIGNAL MICROCONTROLLER

short-form description

CPU

The MSP430 CPU has a 16-bit RISC architecture that is highly transparent to the application. All operations, other than program-flow instructions, are performed as register operations in conjunction with seven addressing modes for source operand and four addressing modes for destination operand.

The CPU is integrated with 16 registers that provide reduced instruction execution time. The register-to-register operation execution time is one cycle of the CPU clock.

Four of the registers, R0 to R3, are dedicated as program counter, stack pointer, status register, and constant generator respectively. The remaining registers are general-purpose registers.

Peripherals are connected to the CPU using data, address, and control buses, and can be handled with all instructions.

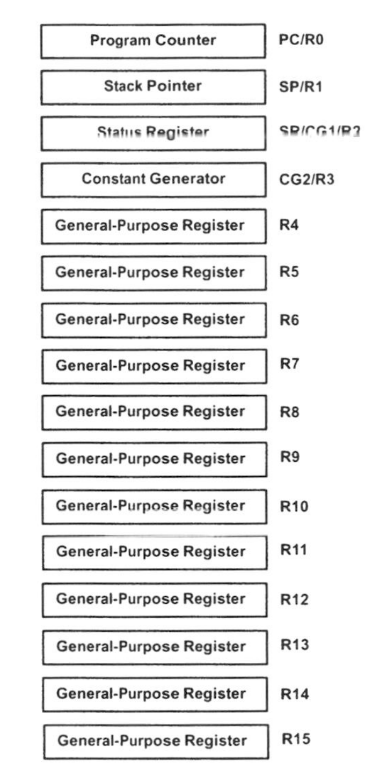

instruction set

The instruction set consists of 51 instructions with three formats and seven address modes. Each instruction can operate on word and byte data. Table 1 shows examples of the three types of instruction formats; the address modes are listed in Table 2.

Table 1. Instruction Word Formats

Dual operands, source-destination	e.g. ADD R4,R5	R4 + R5 —> R5
Single operands, destination only	e.g. CALL R8	PC —>(TOS), R8—> PC
Relative jump, un/conditional	e.g. JNE	Jump-on-equal bit = 0

Table 2. Address Mode Descriptions

ADDRESS MODE	S	D	SYNTAX	EXAMPLE	OPERATION
Register	✓	✓	MOV Rs,Rd	MOV R10,R11	R10 —> R11
Indexed	✓	✓	MOV X(Rn),Y(Rm)	MOV 2(R5),6(R6)	M(2+R5)—> M(6+R6)
Symbolic (PC relative)	✓	✓	MOV EDE,TONI		M(EDE) —> M(TONI)
Absolute	✓	✓	MOV and MEM,and TCDAT		M(MEM) —> M(TCDAT)
Indirect	✓		MOV @Rn,Y(Rm)	MOV @R10,Tab(R6)	M(R10) —> M(Tab+R6)
Indirect autoincrement	✓		MOV @Rn+,Rm	MOV @R10+,R11	M(R10) —> R11 R10 + 2—> R10
Immediate	✓		MOV #X,TONI	MOV #45,TONI	#45 —> M(TONI)

NOTE: S = source D = destination

MSP430x13x, MSP430x14x
MIXED SIGNAL MICROCONTROLLER

operating modes

The MSP430 has one active mode and five software selectable low-power modes of operation. An interrupt event can wake up the device from any of the five low-power modes, service the request and restore back to the low-power mode on return from the interrupt program.

The following six operating modes can be configured by software:

- Active mode AM;
  - All clocks are active
- Low-power mode 0 (LPM0);
  - CPU is disabled
    ACLK and SMCLK remain active. MCLK is disabled
- Low-power mode 1 (LPM1);
  - CPU is disabled
    ACLK and SMCLK remain active. MCLK is disabled
    DCO's dc-generator is disabled if DCO not used in active mode
- Low-power mode 2 (LPM2);
  - CPU is disabled
    MCLK and SMCLK are disabled
    DCO's dc-generator remains enabled
    ACLK remains active
- Low-power mode 3 (LPM3);
  - CPU is disabled
    MCLK and SMCLK are disabled
    DCO's dc-generator is disabled
    ACLK remains active
- Low-power mode 4 (LPM4);
  - CPU is disabled
    ACLK is disabled
    MCLK and SMCLK are disabled
    DCO's dc-generator is disabled
    Crystal oscillator is stopped

MSP430x13x, MSP430x14x
MIXED SIGNAL MICROCONTROLLER

interrupt vector addresses

The interrupt vectors and the power-up starting address are located in the address range 0FFFFh – 0FFE0h. The vector contains the 16-bit address of the appropriate interrupt-handler instruction sequence.

INTERRUPT SOURCE	INTERRUPT FLAG	SYSTEM INTERRUPT	WORD ADDRESS	PRIORITY
Power-up External Reset Watchdog Flash memory	WDTIFG KEYV (see Note 1)	Reset	0FFFEh	15, highest
NMI Oscillator Fault Flash memory access violation	NMIIFG (see Notes 1 & 4) OFIFG (see Notes 1 & 4) ACCVIFG (see Notes 1 & 4)	(Non)maskable (Non)maskable (Non)maskable	0FFFCh	14
Timer_B7 (see Note 5)	TBCCR0 CCIFG (see Note 2)	Maskable	0FFFAh	13
Timer_B7 (see Note 5)	TBCCR1 to 6 CCIFGs, TBIFG (see Notes 1 & 2)	Maskable	0FFF8h	12
Comparator_A	CAIFG	Maskable	0FFF6h	11
Watchdog timer	WDTIFG	Maskable	0FFF4h	10
USART0 receive	URXIFG0	Maskable	0FFF2h	9
USART0 transmit	UTXIFG0	Maskable	0FFF0h	8
ADC12	ADC12IFG (see Notes 1 & 2)	Maskable	0FFEEh	7
Timer_A3	TACCR0 CCIFG (see Note 2)	Maskable	0FFECh	6
Timer_A3	TACCR1 CCIFG, TACCR2 CCIFG, TAIFG (see Notes 1 & 2)	Maskable	0FFEAh	5
I/O port P1 (eight flags)	P1IFG.0 (see Notes 1 & 2) To P1IFG.7 (see Notes 1 & 2)	Maskable	0FFE8h	4
USART1 receive	URXIFG1	Maskable	0FFE6h	3
USART1 transmit	UTXIFG1		0FFE4h	2
I/O port P2 (eight flags)	P2IFG.0 (see Notes 1 & 2) To P2IFG.7 (see Notes 1 & 2)	Maskable	0FFE2h	1
			0FFE0h	0, lowest

NOTES:
1. Multiple source flags
2. Interrupt flags are located in the module.
3. Nonmaskable: neither the individual nor the general interrupt-enable bit will disable an interrupt event.
4. (Non)maskable: the individual interrupt-enable bit can disable an interrupt event, but the general-interrupt enable can not disable it.
5. Timer_B7 in MSP430x14x family has 7 CCRs; Timer_B3 in MSP430x13x family has 3 CCRs. In Timer_B3 there are only interrupt flags TBCCR0, 1, and 2 CCIFGs and the interrupt-enable bits TBCCTL0, 1, and 2 CCIEs.

MSP430x13x, MSP430x14x
MIXED SIGNAL MICROCONTROLLER

special function registers

Most interrupt and module-enable bits are collected in the lowest address space. Special-function register bits not allocated to a functional purpose are not physically present in the device. This arrangement provides simple software access.

interrupt enable 1 and 2

Address	7	6	5	4	3	2	1	0
0h	UTXIE0	URXIE0	ACCVIE	NMIIE			OFIE	WDTIE
	rw-0	rw-0	rw-0	rw-0			rw-0	rw-0

WDTIE: Watchdog-timer interrupt enable. Inactive if watchdog mode is selected. Active if watchdog timer is configured in interval timer mode.

OFIE: Oscillator-fault-interrupt enable

NMIIE: Nonmaskable-interrupt enable

ACCVIE: Flash access violation interrupt enable

URXIE0: USART0, UART, and SPI receive-interrupt enable

UTXIE0: USART0, UART, and SPI transmit-interrupt enable

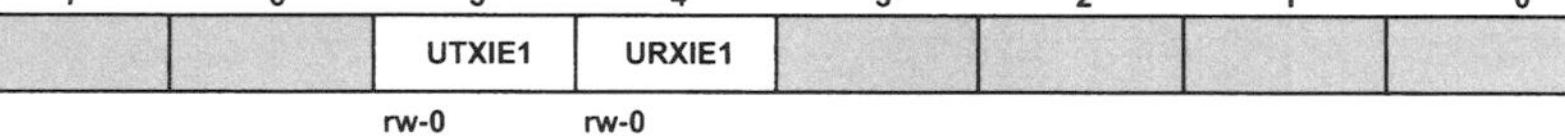

URXIE1: USART1, UART, and SPI receive-interrupt enable

UTXIE1: USART1, UART, and SPI transmit-interrupt enable

interrupt flag register 1 and 2

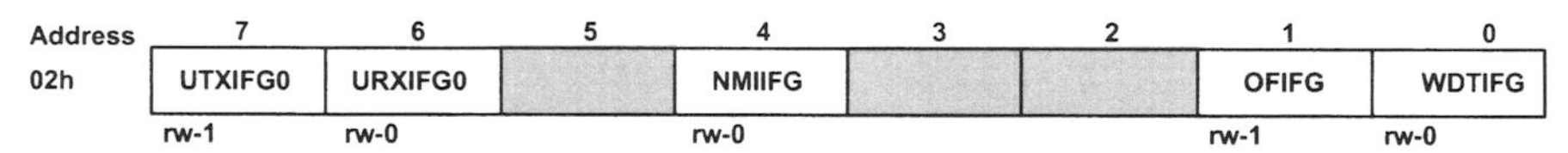

WDTIFG: Set on watchdog timer overflow (in watchdog mode) or security key violation. Reset on V_{CC} power up or a reset condition at the $\overline{RST}$/NMI pin in reset mode.

OFIFG: Flag set on oscillator fault

NMIIFG: Set via $\overline{RST}$/NMI pin

URXIFG0: USART0, UART, and SPI receive flag

UTXIFG0: USART0, UART, and SPI transmit flag

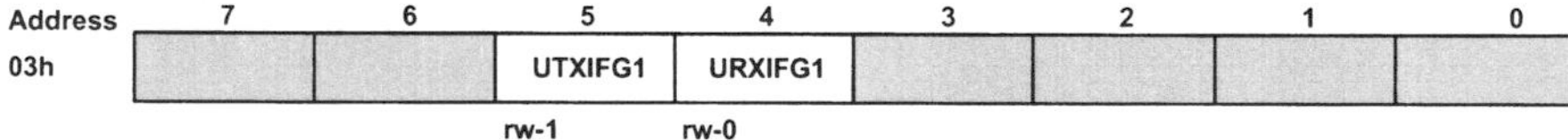

URXIFG1: USART1, UART, and SPI receive flag

UTXIFG1: USART1, UART, and SPI transmit flag

module enable registers 1 and 2

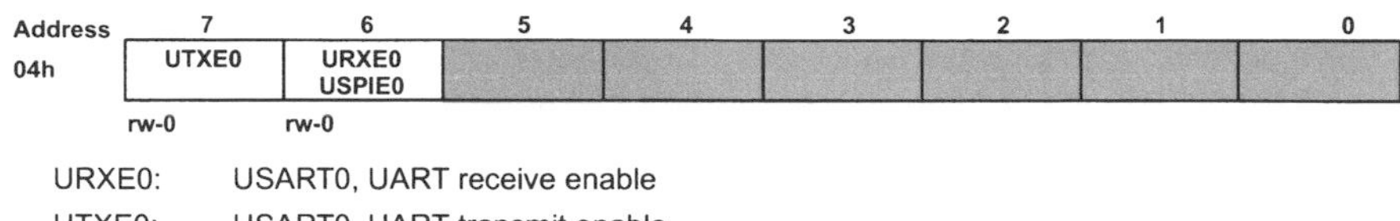

URXE0:	USART0, UART receive enable
UTXE0:	USART0, UART transmit enable
USPIE0:	USART0, SPI (synchronous peripheral interface) transmit and receive enable

Address	7	6	5	4	3	2	1	0
05h			UTXE1	URXE1 USPIE1				
			rw-0	rw-0				

URXE1:	USART1, UART receive enable
UTXE1:	USART1, UART transmit enable
USPIE1:	USART1, SPI (synchronous peripheral interface) transmit and receive enable

Legend: rw: Bit Can Be Read and Written
rw-0: Bit Can Be Read and Written. It Is Reset by PUC.
(shaded): SFR Bit Not Present in Device

memory organization

		MSP430F133	MSP430F135	MSP430F147	MSP430F148	MSP430F149
Memory Main: interrupt vector Main: code memory	Size Flash Flash	8kB 0FFFFh – 0FFE0h 0FFFFh – 0E000h	16kB 0FFFFh – 0FFE0h 0FFFFh – 0C000h	32kB 0FFFFh – 0FFE0h 0FFFFh – 08000h	48kB 0FFFFh – 0FFE0h 0FFFFh – 04000h	60kB 0FFFFh – 0FFE0h 0FFFFh – 01100h
Information memory	Size Flash	256 Byte 010FFh – 01000h	256 Byte 010FFh – 01000h	256 Byte 010FFh – 01000h	256 Byte 010FFh – 01000h	256 Byte 010FFh – 01000h
Boot memory	Size ROM	1kB 0FFFh – 0C00h	1kB 0FFFh – 0C00h	1kB 0FFFh – 0C00h	1kB 0FFFh – 0C00h	1kB 0FFFh – 0C00h
RAM	Size	256 Byte 02FFh – 0200h	512 Byte 03FFh – 0200h	1kB 05FFh – 0200h	2kB 09FFh – 0200h	2kB 09FFh – 0200h
Peripherals	16-bit 8-bit 8-bit SFR	01FFh – 0100h 0FFh – 010h 0Fh – 00h	01FFh – 0100h 0FFh – 010h 0Fh – 00h	01FFh – 0100h 0FFh – 010h 0Fh – 00h	01FFh – 0100h 0FFh – 010h 0Fh – 00h	01FFh – 0100h 0FFh – 010h 0Fh – 00h

bootstrap loader (BSL)

The MSP430 bootstrap loader (BSL) enables users to program the flash memory or RAM using a UART serial interface. Access to the MSP430 memory via the BSL is protected by user-defined password. For complete description of the features of the BSL and its implementation, see the Application report *Features of the MSP430 Bootstrap Loader*, Literature Number SLAA089.

MSP430x13x, MSP430x14x
MIXED SIGNAL MICROCONTROLLER

flash memory

The flash memory can be programmed via the JTAG port, the bootstrap loader, or in-system by the CPU. The CPU can perform single-byte and single-word writes to the flash memory. Features of the flash memory include:

- Flash memory has n segments of main memory and two segments of information memory (A and B) of 128 bytes each. Each segment in main memory is 512 bytes in size.
- Segments 0 to n may be erased in one step, or each segment may be individually erased.
- Segments A and B can be erased individually, or as a group with segments 0–n.
  Segments A and B are also called *information memory.*
- New devices may have some bytes programmed in the information memory (needed for test during manufacturing). The user should perform an erase of the information memory prior to the first use.

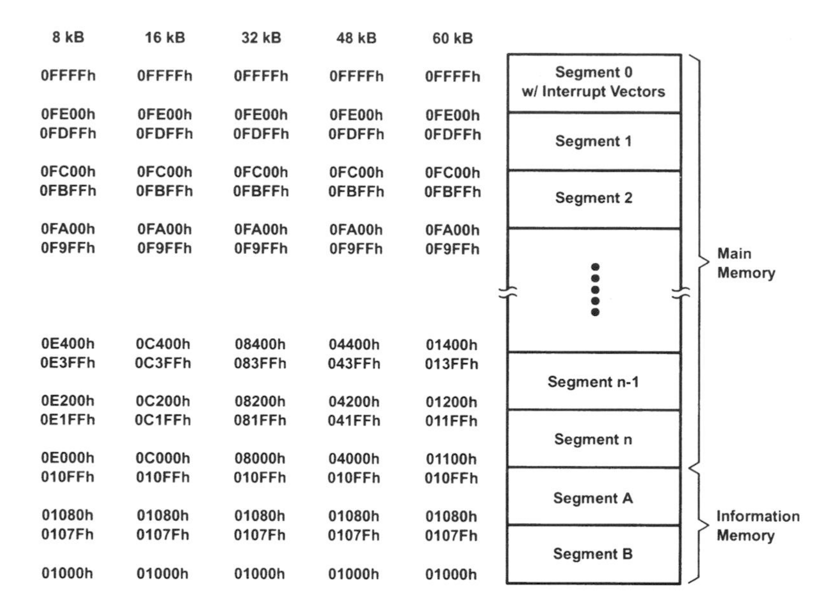

peripherals

Peripherals are connected to the CPU through data, address, and control busses and can be handled using all instructions.

digital I/O

There are six 8-bit I/O ports implemented—ports P1 through P6:

- All individual I/O bits are independently programmable.
- Any combination of input, output, and interrupt conditions is possible.
- Edge-selectable interrupt input capability for all the eight bits of ports P1 and P2.
- Read/write access to port-control registers is supported by all instructions.

oscillator and system clock

The clock system in the MSP430x13x and MSP43x14x family of devices is supported by the basic clock module that includes support for a 32768-Hz watch crystal oscillator, an internal digitally-controlled oscillator (DCO) and a high frequency crystal oscillator. The basic clock module is designed to meet the requirements of both low system cost and low-power consumption. The internal DCO provides a fast turn-on clock source and stabilizes in less than 6 μs. The basic clock module provides the following clock signals:

- Auxiliary clock (ACLK), sourced from a 32768-Hz watch crystal or a high frequency crystal.
- Main clock (MCLK), the system clock used by the CPU.
- Sub-Main clock (SMCLK), the sub-system clock used by the peripheral modules.

watchdog timer

The primary function of the watchdog timer (WDT) module is to perform a controlled system restart after a software problem occurs. If the selected time interval expires, a system reset is generated. If the watchdog function is not needed in an application, the module can be configured as an interval timer and can generate interrupts at selected time intervals.

multiplication (MSP430x14x Only)

The multiplication operation is supported by a dedicated peripheral module. The module performs 16 × 16, 16 × 8, 8 × 16, and 8 × 8 bit operations. The module is capable of supporting signed and unsigned multiplication as well as signed and unsigned multiply and accumulate operations. The result of an operation can be accessed immediately after the operands have been loaded into the peripheral registers. No additional clock cycles are required.

USART0

The MSP430x13x and the MSP430x14x have one hardware universal synchronous/asynchronous receive transmit (USART0) peripheral module that is used for serial data communication. The USART supports synchronous SPI (3 or 4 pin) and asynchronous UART communication protocols, using double-buffered transmit and receive channels.

USART1 (MSP430x14x Only)

The MSP430x14x has a second hardware universal synchronous/asynchronous receive transmit (USART1) peripheral module that is used for serial data communication. The USART supports synchronous SPI (3 or 4 pin) and asynchronous UART communication protocols, using double-buffered transmit and receive channels. Operation of USART1 is identical to USART0.

MSP430x13x, MSP430x14x
MIXED SIGNAL MICROCONTROLLER

timer_A3

Timer_A3 is a 16-bit timer/counter with three capture/compare registers. Timer_A3 can support multiple capture/compares, PWM outputs, and interval timing. Timer_A3 also has extensive interrupt capabilities. Interrupts may be generated from the counter on overflow conditions and from each of the capture/compare registers.

timer_B7 (MSP430x14x Only)

Timer_B7 is a 16-bit timer/counter with seven capture/compare registers. Timer_B7 can support multiple capture/compares, PWM outputs, and interval timing. Timer_B7 also has extensive interrupt capabilities. Interrupts may be generated from the counter on overflow conditions and from each of the capture/compare registers.

timer_B3 (MSP430x13x Only)

Timer_B3 is a 16-bit timer/counter with three capture/compare registers. Timer_B3 can support multiple capture/compares, PWM outputs, and interval timing. Timer_B3 also has extensive interrupt capabilities. Interrupts may be generated from the counter on overflow conditions and from each of the capture/compare registers.

comparator_A

The primary function of the comparator_A module is to support precision slope analog–to–digital conversions, battery–voltage supervision, and monitoring of external analog signals.

ADC12

The ADC12 module supports fast, 12-bit analog-to-digital conversions. The module implements a 12-bit SAR core, sample select control, reference generator and a 16 word conversion-and-control buffer. The conversion-and-control buffer allows up to 16 independent ADC samples to be converted and stored without any CPU intervention.

MSP430x13x, MSP430x14x
MIXED SIGNAL MICROCONTROLLER

peripheral file map

PERIPHERALS WITH WORD ACCESS			
Watchdog	Watchdog Timer control	WDTCTL	0120h
Timer_B7	Timer_B interrupt vector	TBIV	011Eh
Timer_B3	Timer_B control	TBCTL	0180h
(see Note 1)	Capture/compare control 0	TBCCTL0	0182h
	Capture/compare control 1	TBCCTL1	0184h
	Capture/compare control 2	TBCCTL2	0186h
	Capture/compare control 3	TBCCTL3	0188h
	Capture/compare control 4	TBCCTL4	018Ah
	Capture/compare control 5	TBCCTL5	018Ch
	Capture/compare control 6	TBCCTL6	018Eh
	Timer_B register	TBR	0190h
	Capture/compare register 0	TBCCR0	0192h
	Capture/compare register 1	TBCCR1	0194h
	Capture/compare register 2	TBCCR2	0196h
	Capture/compare register 3	TBCCR3	0198h
	Capture/compare register 4	TBCCR4	019Ah
	Capture/compare register 5	TBCCR5	019Ch
	Capture/compare register 6	TBCCR6	019Eh
Timer_A3	Timer_A interrupt vector	TAIV	012Eh
	Timer_A control	TACTL	0160h
	Capture/compare control 0	TACCTL0	0162h
	Capture/compare control 1	TACCTL1	0164h
	Capture/compare control 2	TACCTL2	0166h
	Reserved		0168h
	Reserved		016Ah
	Reserved		016Ch
	Reserved		016Eh
	Timer_A register	TAR	0170h
	Capture/compare register 0	TACCR0	0172h
	Capture/compare register 1	TACCR1	0174h
	Capture/compare register 2	TACCR2	0176h
	Reserved		0178h
	Reserved		017Ah
	Reserved		017Ch
	Reserved		017Eh
Multiply	Sum extend	SUMEXT	013Eh
(MSP430x14x only)	Result high word	RESHI	013Ch
	Result low word	RESLO	013Ah
	Second operand	OP2	0138h
	Multiply signed +accumulate/operand1	MACS	0136h
	Multiply+accumulate/operand1	MAC	0134h
	Multiply signed/operand1	MPYS	0132h
	Multiply unsigned/operand1	MPY	0130h

NOTE 1: Timer_B7 in MSP430x14x family has 7 CCR, Timer_B3 in MSP430x13x family has 3 CCR.

MSP430x13x, MSP430x14x
MIXED SIGNAL MICROCONTROLLER

peripheral file map (continued)

PERIPHERALS WITH WORD ACCESS (CONTINUED)			
Flash	Flash control 3	FCTL3	012Ch
	Flash control 2	FCTL2	012Ah
	Flash control 1	FCTL1	0128h
ADC12	Conversion memory 15	ADC12MEM15	015Eh
	Conversion memory 14	ADC12MEM14	015Ch
	Conversion memory 13	ADC12MEM13	015Ah
	Conversion memory 12	ADC12MEM12	0158h
	Conversion memory 11	ADC12MEM11	0156h
	Conversion memory 10	ADC12MEM10	0154h
	Conversion memory 9	ADC12MEM9	0152h
	Conversion memory 8	ADC12MEM8	0150h
	Conversion memory 7	ADC12MEM7	014Eh
	Conversion memory 6	ADC12MEM6	014Ch
	Conversion memory 5	ADC12MEM5	014Ah
	Conversion memory 4	ADC12MEM4	0148h
	Conversion memory 3	ADC12MEM3	0146h
	Conversion memory 2	ADC12MEM2	0144h
	Conversion memory 1	ADC12MEM1	0142h
	Conversion memory 0	ADC12MEM0	0140h
	Interrupt-vector-word register	ADC12IV	01A8h
	Inerrupt-enable register	ADC12IE	01A6h
	Inerrupt-flag register	ADC12IFG	01A4h
	Control register 1	ADC12CTL1	01A2h
	Control register 0	ADC12CTL0	01A0h
	ADC memory-control register15	ADC12MCTL15	08Fh
	ADC memory-control register14	ADC12MCTL14	08Eh
	ADC memory-control register13	ADC12MCTL13	08Dh
	ADC memory-control register12	ADC12MCTL12	08Ch
	ADC memory-control register11	ADC12MCTL11	08Bh
	ADC memory-control register10	ADC12MCTL10	08Ah
	ADC memory-control register9	ADC12MCTL9	089h
	ADC memory-control register8	ADC12MCTL8	088h
	ADC memory-control register7	ADC12MCTL7	087h
	ADC memory-control register6	ADC12MCTL6	086h
	ADC memory-control register5	ADC12MCTL5	085h
	ADC memory-control register4	ADC12MCTL4	084h
	ADC memory-control register3	ADC12MCTL3	083h
	ADC memory-control register2	ADC12MCTL2	082h
	ADC memory-control register1	ADC12MCTL1	081h
	ADC memory-control register0	ADC12MCTL0	080h

peripheral file map (continued)

PERIPHERALS WITH BYTE ACCESS			
USART1 (Only in 'x14x)	Transmit buffer	U1TXBUF	07Fh
	Receive buffer	U1RXBUF	07Eh
	Baud rate	U1BR1	07Dh
	Baud rate	U1BR0	07Ch
	Modulation control	U1MCTL	07Bh
	Receive control	U1RCTL	07Ah
	Transmit control	U1TCTL	079h
	USART control	U1CTL	078h
USART0	Transmit buffer	U0TXBUF	077h
	Receive buffer	U0RXBUF	076h
	Baud rate	U0BR1	075h
	Baud rate	U0BR0	074h
	Modulation control	U0MCTL	073h
	Receive control	U0RCTL	072h
	Transmit control	U0TCTL	071h
	USART control	U0CTL	070h
Comparator_A	Comparator_A port disable	CAPD	05Bh
	Comparator_A control2	CACTL2	05Ah
	Comparator_A control1	CACTL1	059h
Basic Clock	Basic clock system control2	BCSCTL2	058h
	Basic clock system control1	BCSCTL1	057h
	DCO clock frequency control	DCOCTL	056h
Port P6	Port P6 selection	P6SEL	037h
	Port P6 direction	P6DIR	036h
	Port P6 output	P6OUT	035h
	Port P6 input	P6IN	034h
Port P5	Port P5 selection	P5SEL	033h
	Port P5 direction	P5DIR	032h
	Port P5 output	P5OUT	031h
	Port P5 input	P5IN	030h
Port P4	Port P4 selection	P4SEL	01Fh
	Port P4 direction	P4DIR	01Eh
	Port P4 output	P4OUT	01Dh
	Port P4 input	P4IN	01Ch
Port P3	Port P3 selection	P3SEL	01Bh
	Port P3 direction	P3DIR	01Ah
	Port P3 output	P3OUT	019h
	Port P3 input	P3IN	018h
Port P2	Port P2 selection	P2SEL	02Eh
	Port P2 interrupt enable	P2IE	02Dh
	Port P2 interrupt-edge select	P2IES	02Ch
	Port P2 interrupt flag	P2IFG	02Bh
	Port P2 direction	P2DIR	02Ah
	Port P2 output	P2OUT	029h
	Port P2 input	P2IN	028h

MSP430x13x, MSP430x14x
MIXED SIGNAL MICROCONTROLLER

peripheral file map (continued)

PERIPHERALS WITH BYTE ACCESS			
Port P1	Port P1 selection	P1SEL	026h
	Port P1 interrupt enable	P1IE	025h
	Port P1 interrupt-edge select	P1IES	024h
	Port P1 interrupt flag	P1IFG	023h
	Port P1 direction	P1DIR	022h
	Port P1 output	P1OUT	021h
	Port P1 input	P1IN	020h
Special Functions	SFR module enable 2	ME2	005h
	SFR module enable 1	ME1	004h
	SFR interrupt flag2	IFG2	003h
	SFR interrupt flag1	IFG1	002h
	SFR interrupt enable2	IE2	001h
	SFR interrupt enable1	IE1	000h

absolute maximum ratings over operating free-air temperature (unless otherwise noted)†

Voltage applied at V_{CC} to V_{SS} . . . . −0.3 V to + 4.1 V
Voltage applied to any pin (referenced to V_{SS}) . . . . −0.3 V to V_{CC}+0.3 V
Diode current at any device terminal . . . . ±2 mA
Storage temperature (unprogrammed device) . . . . −55°C to 150°C
Storage temperature (programmed device) . . . . −40°C to 85°C

† Stresses beyond those listed under "absolute maximum ratings" may cause permanent damage to the device. These are stress ratings only, and functional operation of the device at these or any other conditions beyond those indicated under "recommended operating conditions" is not implied. Exposure to absolute-maximum-rated conditions for extended periods may affect device reliability.

NOTE: All voltages referenced to V_{SS}.

MSP430x13x, MSP430x14x
MIXED SIGNAL MICROCONTROLLER

recommended operating conditions

PARAMETER			MIN	NOM	MAX	UNITS
Supply voltage during program execution, V_{CC} ($AV_{CC} = DV_{CC} = V_{CC}$)		MSP430F13x, MSP430F14x	1.8		3.6	V
Supply voltage during flash memory programming, V_{CC} ($AV_{CC} = DV_{CC} = V_{CC}$)		MSP430F13x, MSP430F14x	2.7		3.6	V
Supply voltage, V_{SS} ($AV_{SS} = DV_{SS} = V_{SS}$)			0.0		0.0	V
Operating free-air temperature range, T_A		MSP430x13x MSP430x14x	–40		85	°C
LFXT1 crystal frequency, $f_{(LFXT1)}$ (see Notes 1 and 2)	LF selected, XTS=0	Watch crystal		32768		Hz
	XT1 selected, XTS=1	Ceramic resonator	450		8000	kHz
	XT1 selected, XTS=1	Crystal	1000		8000	kHz
XT2 crystal frequency, $f_{(XT2)}$		Ceramic resonator	450		8000	kHz
		Crystal	1000		8000	
Processor frequency (signal MCLK), $f_{(System)}$		V_{CC} = 1.8 V	DC		4.15	MHz
		V_{CC} = 3.6 V	DC		8	
Flash-timing-generator frequency, $f_{(FTG)}$		MSP430F13x, MSP430F14x	257		476	kHz
Cumulative program time, $t_{(CPT)}$ (see Note 3)		V_{CC} = 2.7 V/3.6 V MSP430F13x MSP430F14x			3	ms
Mass erase time, t(MEras) (See also the *flash memory, timing generator, control register FCTL2* section, see Note 4)		V_{CC} = 2.7 V/3.6 V	200			ms
Low-level input voltage (TCK, TMS, TDI, RST/NMI), V_{IL} (excluding Xin, Xout)		V_{CC} = 2.2 V/3 V	V_{SS}		V_{SS} +0.6	V
High-level input voltage (TCK, TMS, TDI, RST/NMI), V_{IH} (excluding Xin, Xout)		V_{CC} = 2.2 V/3 V	$0.8V_{CC}$		V_{CC}	V
Input levels at Xin and Xout	$V_{IL(Xin, Xout)}$	V_{CC} = 2.2 V/3 V	V_{SS}		$0.2 \times V_{CC}$	V
	$V_{IH(Xin, Xout)}$		$0.8 \times V_{CC}$		V_{CC}	

NOTES: 1. In LF mode, the LFXT1 oscillator requires a watch crystal and the LFXT1 oscillator requires a 5.1-MΩ resistor from XOUT to V_{SS} when V_{CC} < 2.5 V. In XT1 mode, the LFXT1. and XT2 oscillators accept a ceramic resonator or a 4-MHz crystal frequency at $V_{CC} \geq 2.2$ V. In XT1 mode, the LFXT1 and XT2 oscillators accept a ceramic resonator or an 8-MHz crystal frequency at V_{CC} > 2.8 V.
2. In LF mode, the LFXT1 oscillator requires a watch crystal. In XT1 mode, FXT1 accepts a ceramic resonator or a crystal.
3. The cumulative program time must not be exceeded during a block-write operation. This parameter is only relevant if segment write option is used.
4. The mass erase duration generated by the flash timing generator is at least 11.1 ms. The cummulative mass erase time needed is 200 ms. This can be achieved by repeating the mass erase operation until the cumulative mass erase time is met (a minimum of 19 cycles may be required).

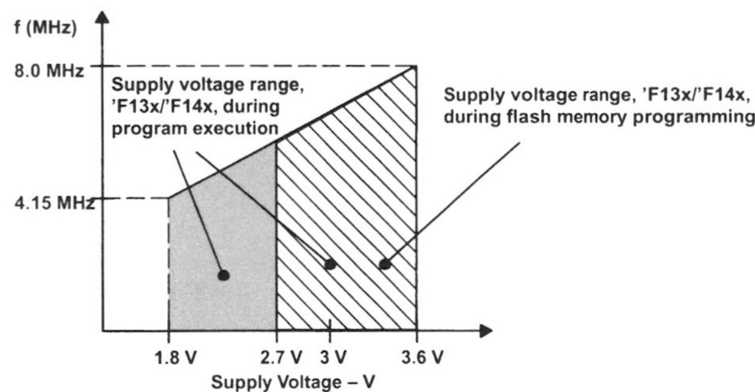

Figure 1. Frequency vs Supply Voltage, MSP430F13x or MSP430F14x

MSP430x13x, MSP430x14x
MIXED SIGNAL MICROCONTROLLER

electrical characteristics over recommended operating free-air temperature (unless otherwise noted)

supply current into AV_{CC} + DV_{CC} excluding external current

PARAMETER			TEST CONDITIONS		MIN	NOM	MAX	UNIT
$I_{(AM)}$	Active mode, (see Note 1) $f_{(MCLK)} = f_{(SMCLK)}$ = 1 MHz, $f_{(ACLK)}$ = 32,768 Hz XTS=0, SELM=(0,1)	F13x, F14x	T_A = –40°C to 85°C	V_{CC} = 2.2 V		280	350	μA
				V_{CC} = 3 V		420	560	
$I_{(AM)}$	Active mode, (see Note 1) $f_{(MCLK)} = f_{(SMCLK)}$ = 4 096 Hz, $f_{(ACLK)}$ = 4,096 Hz XTS=0, SELM=(0,1) XTS=0, SELM=3	F13x, F14x	T_A = –40°C to 85°C	V_{CC} = 2.2 V		2.5	7	μA
				V_{CC} = 3 V		9	20	
$I_{(LPM0)}$	Low-power mode, (LPM0) (see Note 1)	F13x, F14x	T_A = –40°C to 85°C	V_{CC} = 2.2 V		32	45	μA
				V_{CC} = 3 V		55	70	
$I_{(LPM2)}$	Low-power mode, (LPM2), f(MCLK) = f (SMCLK) = 0 MHz, f(ACLK) = 32.768 Hz, SCG0 = 0		T_A = –40°C to 85°C	V_{CC} = 2.2 V		11	14	μA
				V_{CC} = 3 V		17	22	
$I_{(LPM3)}$	Low-power mode, (LPM3) $f_{(MCLK)} = f_{(SMCLK)}$ = 0 MHz, $f_{(ACLK)}$ = 32,768 Hz, SCG0 = 1 (see Note 2)		T_A = –40°C	V_{CC} = 2.2 V		0.8	1.5	μA
			T_A = 25°C			0.9	1.5	
			T_A = 85°C			1.6	2.8	
			T_A = –40°C	V_{CC} = 3 V		1.8	2.2	μA
			T_A = 25°C			1.6	1.9	
			T_A = 85°C			2.3	3.9	
$I_{(LPM4)}$	Low-power mode, (LPM4) $f_{(MCLK)}$ = 0 MHz, $f_{(SMCLK)}$ = 0 MHz, $f_{(ACLK)}$ = 0 Hz, SCG0 = 1		T_A = –40°C	V_{CC} = 2.2 V		0.1	0.5	μA
			T_A = 25°C			0.1	0.5	
			T_A = 85°C			0.8	2.5	
			T_A = –40°C	V_{CC} = 3 V		0.1	0.5	μA
			T_A = 25°C			0.1	0.5	
			T_A = 85°C			0.8	2.5	

NOTES: 1. Timer_B is clocked by f(DCOCLK) = 1 MHz. All inputs are tied to 0 V or to V_{CC}. Outputs do not source or sink any current.
2. Timer_B is clocked by f(ACLK) = 32,768 Hz. All inputs are tied to 0 V or to V_{CC}. Outputs do not source or sink any current. The current consumption in LPM2 and LPM3 are measured with ACLK selected.

MSP430x13x, MSP430x14x
MIXED SIGNAL MICROCONTROLLER

electrical characteristics over recommended operating free-air temperature (unless otherwise noted) (continued)

Current consumption of active mode versus system frequency, F-version

$I_{(AM)} = I_{(AM)} [1\ MHz] \times f_{(System)} [MHz]$

Current consumption of active mode versus supply voltage, F-version

$I_{(AM)} = I_{(AM)} [3\ V] + 175\ \mu A/V \times (V_{CC} - 3\ V)$

SCHMITT-trigger inputs – Ports P1, P2, P3, P4, P5, and P6

PARAMETER		TEST CONDITIONS	MIN	TYP	MAX	UNIT
V_{IT+}	Positive-going input threshold voltage	V_{CC} = 2.2 V	1.1		1.5	V
		V_{CC} = 3 V	1.5		1.9	
V_{IT-}	Negative-going input threshold voltage	V_{CC} = 2.2 V	0.4		0.9	V
		V_{CC} = 3 V	0.90		1.3	
V_{hys}	Input voltage hysteresis ($V_{IT+} - V_{IT-}$)	V_{CC} = 2.2 V	0.3		1.1	V
		V_{CC} = 3 V	0.5		1	

standard inputs – RST/NMI; JTAG: TCK, TMS, TDI, TDO/TDI

PARAMETER		TEST CONDITIONS	MIN	TYP	MAX	UNIT
V_{IL}	Low-level input voltage	V_{CC} = 2.2 V / 3 V	V_{SS}		V_{SS}+0.6	V
V_{IH}	High-level input voltage		0.8×V_{CC}		V_{CC}	V

outputs – Ports P1, P2, P3, P4, P5, and P6

PARAMETER		TEST CONDITIONS			MIN	TYP	MAX	UNIT
V_{OH}	High-level output voltage	$I_{OH(max)}$ = –1 mA,	V_{CC} = 2.2 V,	See Note 1	V_{CC}–0.25		V_{CC}	V
		$I_{OH(max)}$ = –3.4 mA,	V_{CC} = 2.2 V,	See Note 2	V_{CC}–0.6		V_{CC}	
		$I_{OH(max)}$ = –1 mA,	V_{CC} = 3 V,	See Note 1	V_{CC}–0.25		V_{CC}	
		$I_{OH(max)}$ = –3.4 mA,	V_{CC} = 3 V,	See Note 2	V_{CC}–0.6		V_{CC}	
V_{OL}	Low-level output voltage	$I_{OL(max)}$ = 1.5 mA,	V_{CC} = 2.2 V,	See Note 1	V_{SS}		V_{SS}+0.25	V
		$I_{OL(max)}$ = 6 mA,	V_{CC} = 2.2 V,	See Note 2	V_{SS}		V_{SS}+0.6	
		$I_{OL(max)}$ = 1.5 mA,	V_{CC} = 3 V,	See Note 1	V_{SS}		V_{SS}+0.25	
		$I_{OL(max)}$ = 6 mA,	V_{CC} = 3 V,	See Note 2	V_{SS}		V_{SS}+0.6	

NOTES: 1. The maximum total current, $I_{OH(max)}$ and $I_{OL(max)}$, for all outputs combined, should not exceed ±6 mA to satisfy the maximum specified voltage drop.
2. The maximum total current, $I_{OH(max)}$ and $I_{OL(max)}$, for all outputs combined, should not exceed ±24 mA to satisfy the maximum specified voltage drop.

MSP430x13x, MSP430x14x
MIXED SIGNAL MICROCONTROLLER

outputs – Ports P1, P2, P3, P4, P5, and P6 (continued)

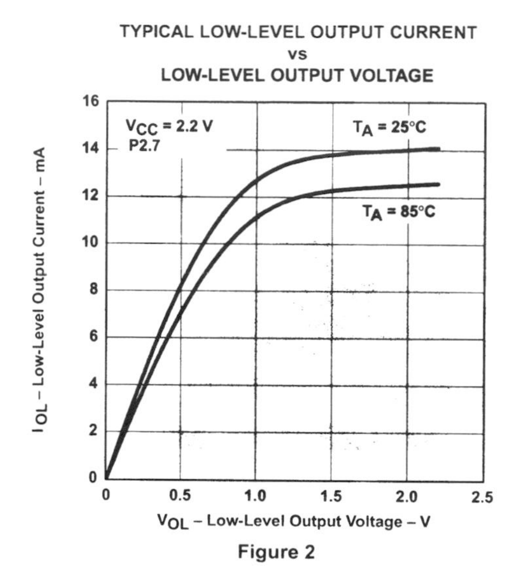

Figure 2

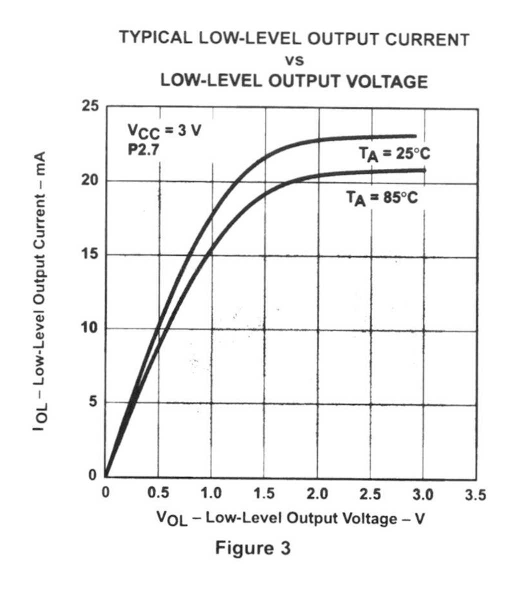

Figure 3

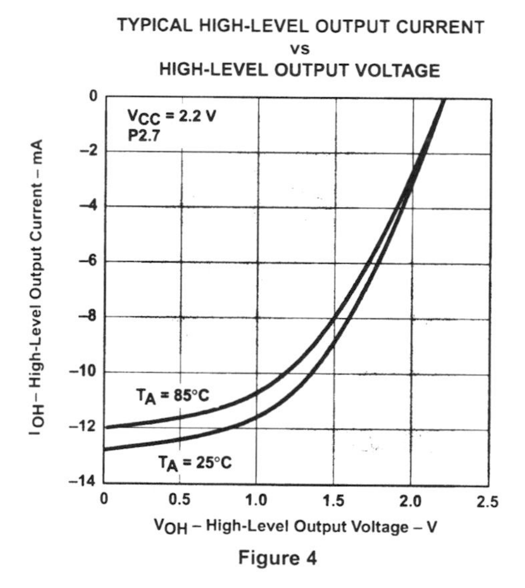

Figure 4

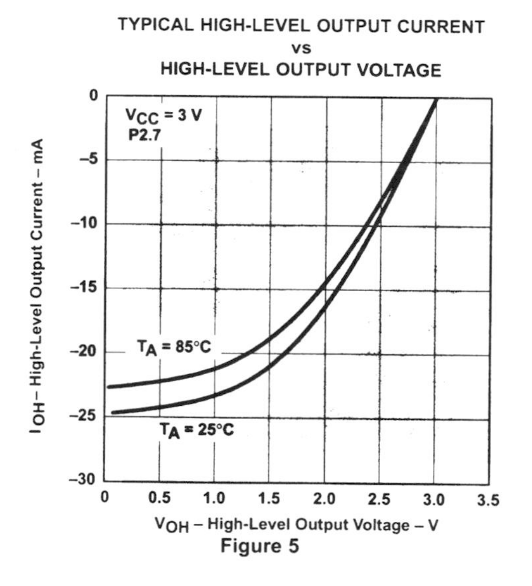

Figure 5

MSP430x13x, MSP430x14x MIXED SIGNAL MICROCONTROLLER

electrical characteristics over recommended operating free-air temperature (unless otherwise noted) (continued)

output frequency

PARAMETER		TEST CONDITIONS		MIN	TYP	MAX	UNIT
f_{TAx}	TA0..2, TB0–TB6, Internal clock source, SMCLK signal applied (see Note 1)	C_L = 20 pF		DC		f_{System}	MHz
f_{ACLK}, f_{MCLK}, f_{SMCLK}	P5.6/ACLK, P5.4/MCLK, P5.5/SMCLK	C_L = 20 pF				f_{System}	MHz
t_{Xdc}	Duty cycle of output frequency,	P2.0/ACLK C_L = 20 pF, V_{CC} = 2.2 V / 3 V	$f_{ACLK} = f_{LFXT1} = f_{XT1}$	40%		60%	
			$f_{ACLK} = f_{LFXT1} = f_{LF}$	30%		70%	
			$f_{ACLK} = f_{LFXT1/n}$		50%		
		P1.4/SMCLK, C_L = 20 pF, V_{CC} = 2.2 V / 3 V	$f_{SMCLK} = f_{LFXT1} = f_{XT1}$	40%		60%	
			$f_{SMCLK} = f_{LFXT1} = f_{LF}$	35%		65%	
			$f_{SMCLK} = f_{LFXT1/n}$	50%– 15 ns	50%	50%– 15 ns	
			$f_{SMCLK} = f_{DCOCLK}$	50%– 15 ns	50%	50%– 15 ns	

NOTE 1: The limits of the system clock MCLK has to be met; the system (MCLK) frequency should not exceed the limits. MCLK and SMCLK frequencies can be different.

inputs Px.x, TAx, TBx

PARAMETER		TEST CONDITIONS	V_{CC}	MIN	TYP	MAX	UNIT
$t_{(int)}$	External interrupt timing	Port P1, P2: P1.x to P2.x, external trigger signal for the interrupt flag, (see Note 1)	2.2 V/3 V	1.5			cycle
			2.2 V	62			ns
			3 V	50			ns
$t_{(cap)}$	Timer_A, Timer_B capture timing	TA0, TA1, TA2 (see Note 2)	2.2 V/3 V	1.5			cycle
			2.2 V	62			ns
		TB0, TB1, TB2, TB3, TB4, TB5, TB6 (see Note 3)	3 V	50			ns
$f_{(TAext)}$	Timer_A, Timer_B clock frequency externally applied to pin	TACLK, TBCLK, INCLK: $t_{(H)} = t_{(L)}$	2.2 V			8	MHz
$f_{(TBext)}$			3 V			10	MHz
$f_{(TAint)}$	Timer_A, Timer_B clock frequency	SMCLK or ACLK signal selected	2.2 V			8	MHz
$f_{(BTAint)}$			3 V			10	MHz

NOTES: 1. The external signal sets the interrupt flag every time the minimum $t_{(int)}$ cycle and time parameters are met. It may be set even with trigger signals shorter than $t_{(int)}$. Both the cycle and timing specifications must be met to ensure the flag is set. $t_{(int)}$ is measured in MCLK cycles.

2. The external capture signal triggers the capture event every time the minimum $t_{(cap)}$ cycle and time parameters are met. A capture may be triggered with capture signals even shorter than $t_{(cap)}$. Both the cycle and timing specifications must be met to ensure a correct capture of the 16-bit timer value and to ensure the flag is set.

3. Seven capture/compare registers in 'x14x and three capture/compare registers in 'x13x.

wake-up LPM3

PARAMETER		TEST CONDITIONS		MIN	TYP	MAX	UNIT
$t_{(LPM3)}$	Delay time	f = 1 MHz	V_{CC} = 2.2 V/3 V			6	µs
		f = 2 MHz				6	µs
		f = 3 MHz				6	µs

MSP430x13x, MSP430x14x
MIXED SIGNAL MICROCONTROLLER

electrical characteristics over recommended operating free-air temperature (unless otherwise noted) (continued)

leakage current (see Note 1)

PARAMETER			TEST CONDITIONS		MIN	TYP	MAX	UNIT
$I_{lkg(P1.x)}$	Leakage current (see Note 1)	Port P1	Port 1: $V_{(P1.x)}$ (see Note 2)	V_{CC} = 2.2 V/3 V			±50	nA
$I_{lkg(P2.x)}$		Port P2	Port 2: $V_{(P2.3)}$ $V_{(P2.4)}$ (see Note 2)				±50	
$I_{lkg(P6.x)}$		Port P6	Port 6: $V_{(P6.x)}$ (see Note 2)				±50	

NOTES: 1. The leakage current is measured with V_{SS} or V_{CC} applied to the corresponding pin(s), unless otherwise noted.
2. The port pin must be selected as input and there must be no optional pullup or pulldown resistor.

RAM

PARAMETER	TEST CONDITIONS	MIN	TYP	MAX	UNIT
VRAMh	CPU HALTED (see Note 1)	1.6			V

NOTE 1: This parameter defines the minimum supply voltage when the data in program memory RAM remain unchanged. No program execution should take place during this supply voltage condition.

Comparator_A (see Note 1)

PARAMETER		TEST CONDITIONS		MIN	TYP	MAX	UNIT
$I_{(DD)}$		CAON=1, CARSEL=0, CAREF=0	V_{CC} = 2.2 V		25	40	μA
			V_{CC} = 3 V		45	60	
$I_{(Refladder/Refdiode)}$		CAON=1, CARSEL=0, CAREF=1/2/3, no load at P2.3/CA0/TA1 and P2.4/CA1/TA2	V_{CC} = 2.2 V		30	50	μA
			V_{CC} = 3 V		45	71	
$V_{(IC)}$	Common-mode input voltage	CAON =1	V_{CC} = 2.2 V/3 V	0		V_{CC}–1	V
$V_{(Ref025)}$ See Figure 6	$\frac{\text{Voltage @ 0.25 } V_{CC} \text{ node}}{V_{CC}}$	PCA0=1, CARSEL=1, CAREF=1, no load at P2.3/CA0/TA1 and P2.4/CA1/TA2, See Figure 6	V_{CC} = 2.2 V/3 V	0.23	0.24	0.25	
$V_{(Ref050)}$ See Figure 6	$\frac{\text{Voltage @ 0.5 } V_{CC} \text{ node}}{V_{CC}}$	PCA0=1, CARSEL=1, CAREF=2, no load at P2.3/CA0/TA1 and P2.4/CA1/TA2, See Figure 6	V_{CC} = 2.2 V/3 V	0.47	0.48	0.5	
$V_{(RefVT)}$		PCA0=1, CARSEL=1, CAREF=3, no load at P2.3/CA0/TA1 and P2.4/CA1/TA2 T_A = 85°C	V_{CC} = 2.2 V	390	480	540	mV
			V_{CC} = 3 V	400	490	550	
$V_{(offset)}$	Offset voltage	See Note 2	V_{CC} = 2.2 V/3 V	–30		30	mV
V_{hys}	Input hysteresis	CAON=1	V_{CC} = 2.2 V/3 V	0	0.7	1.4	mV
$t_{(response\ LH)}$		T_A = 25°C, Overdrive 10 mV, Without filter: CAF=0	V_{CC} = 2.2 V	130	210	300	ns
			V_{CC} = 3 V	80	150	240	
		T_A = 25°C, Overdrive 10 mV, With filter: CAF=1	V_{CC} = 2.2 V	1.4	1.9	3.4	μs
			V_{CC} = 3 V	0.9	1.5	2.6	
$t_{(response\ HL)}$		T_A = 25°C, Overdrive 10 mV, without filter: CAF=0	V_{CC} = 2.2 V	130	210	300	ns
			V_{CC} = 3 V	80	150	240	
		T_A = 25°C, Overdrive 10 mV, with filter: CAF=1	V_{CC} = 2.2 V	1.4	1.9	3.4	μs
			V_{CC} = 3 V	0.9	1.5	2.6	

NOTES: 1. The leakage current for the Comparator_A terminals is identical to $I_{lkg(Px.x)}$ specification.
2. The input offset voltage can be cancelled by using the CAEX bit to invert the Comparator_A inputs on successive measurements. The two successive measurements are then summed together.

MSP430x13x, MSP430x14x
MIXED SIGNAL MICROCONTROLLER

electrical characteristics over recommended operating free-air temperature (unless otherwise noted) (continued)

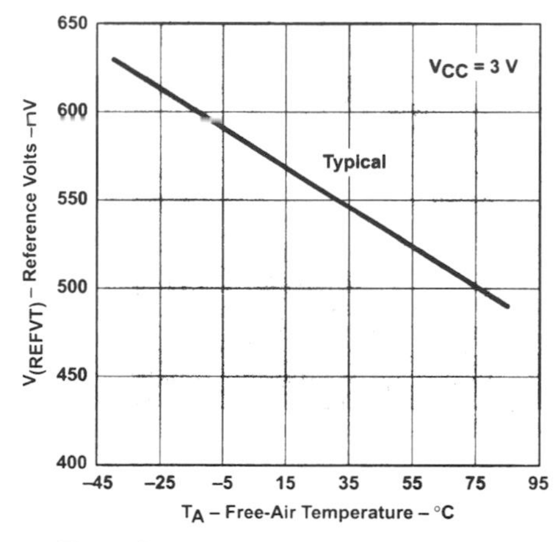

Figure 6. $V_{(RefVT)}$ vs Temperature, V_{CC} = 3 V

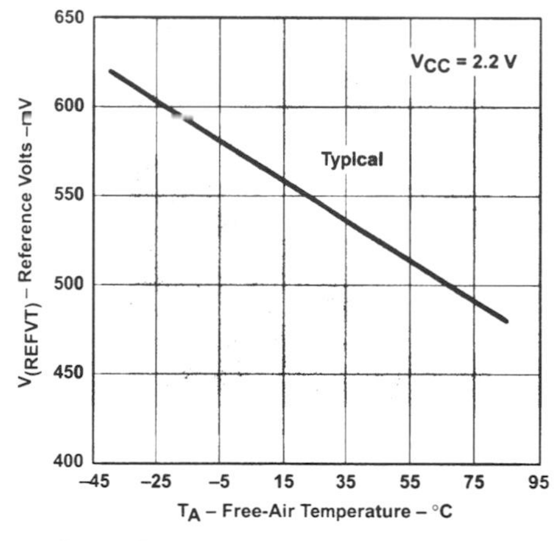

Figure 7. $V_{(RefVT)}$ vs Temperature, V_{CC} = 2.2 V

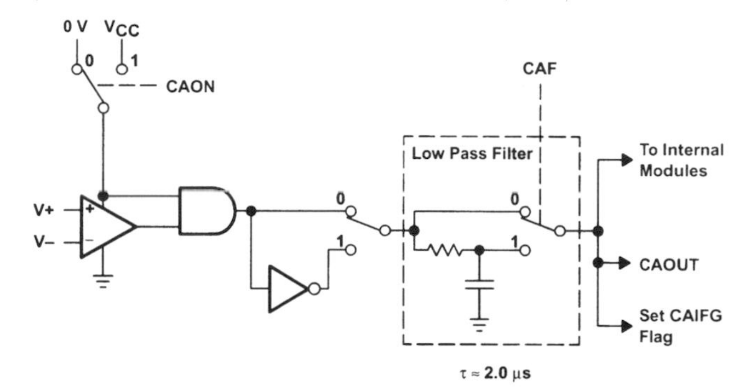

Figure 8. Block Diagram of Comparator_A Module

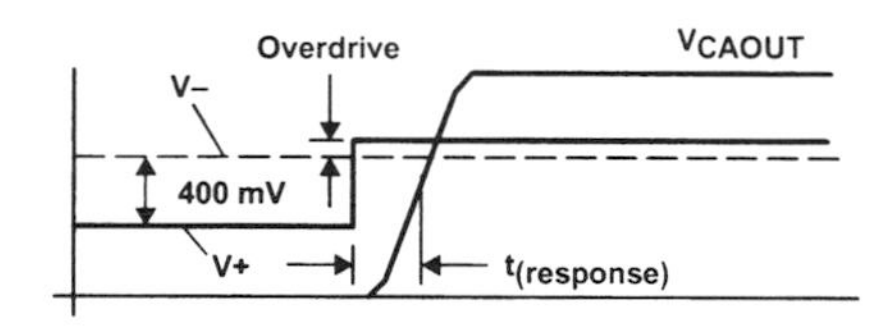

Figure 9. Overdrive Definition

MSP430x13x, MSP430x14x
MIXED SIGNAL MICROCONTROLLER

electrical characteristics over recommended operating free-air temperature (unless otherwise noted) (continued)

POR

PARAMETER		CONDITIONS	V_{CC}	MIN	NOM	MAX	UNIT
$t_{(POR\_Delay)}$	POR		2.2 V/3 V		150	250	μs
V_{POR}		T_A = –40°C		1.4		1.8	V
V_{POR}		T_A = +25°C		1.1		1.5	V
V_{POR}		T_A = +85°C		0.8		1.2	V
$V_{(min)}$				0		0.4	V
$t_{(Reset)}$	PUC/POR	Reset is accepted internally	2.2 V/3 V	2			μs

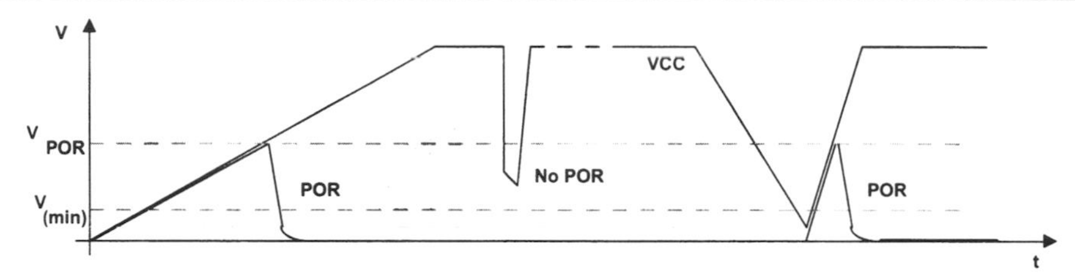

Figure 10. Power-On Reset (POR) vs Supply Voltage

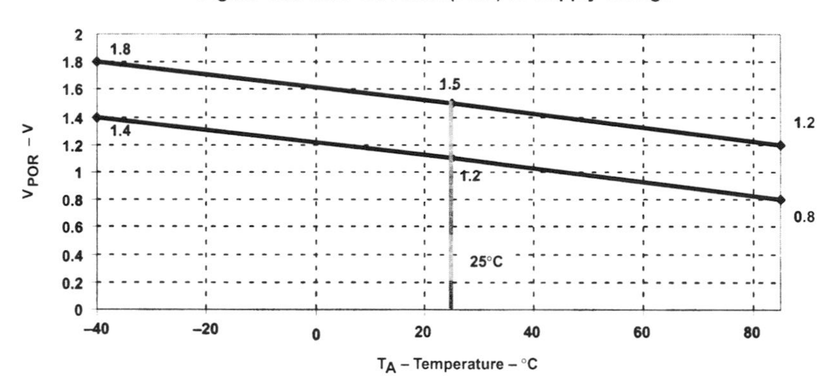

Figure 11. V_{POR} vs Temperature

MSP430x13x, MSP430x14x
MIXED SIGNAL MICROCONTROLLER

electrical characteristics over recommended operating free-air temperature (unless otherwise noted) (continued)

DCO (see Note 1)

PARAMETER	TEST CONDITIONS		MIN	NOM	MAX	UNIT
$f_{(DCO03)}$	R_{sel} = 0, DCO = 3, MOD = 0, DCOR = 0, T_A = 25°C	V_{CC} = 2.2 V	0.08	0.12	0.15	MHz
		V_{CC} = 3 V	0.08	0.13	0.16	
$f_{(DCO13)}$	R_{sel} = 1, DCO = 3, MOD = 0, DCOR = 0, T_A = 25°C	V_{CC} = 2.2 V	0.14	0.19	0.23	MHz
		V_{CC} = 3 V	0.14	0.18	0.22	
$f_{(DCO23)}$	R_{sel} = 2, DCO = 3, MOD = 0, DCOR = 0, T_A = 25°C	V_{CC} = 2.2 V	0.22	0.30	0.36	MHz
		V_{CC} = 3 V	0.22	0.28	0.34	
$f_{(DCO33)}$	R_{sel} = 3, DCO = 3, MOD = 0, DCOR = 0, T_A = 25°C	V_{CC} = 2.2 V	0.37	0.49	0.59	MHz
		V_{CC} = 3 V	0.37	0.47	0.56	
$f_{(DCO43)}$	R_{sel} = 4, DCO = 3, MOD = 0, DCOR = 0, T_A = 25°C	V_{CC} = 2.2 V	0.61	0.77	0.93	MHz
		V_{CC} = 3 V	0.61	0.75	0.90	
$f_{(DCO53)}$	R_{sel} = 5, DCO = 3, MOD = 0, DCOR = 0, T_A = 25°C	V_{CC} = 2.2 V	1	1.2	1.5	MHz
		V_{CC} = 3 V	1	1.3	1.5	
$f_{(DCO63)}$	R_{sel} = 6, DCO = 3, MOD = 0, DCOR = 0, T_A = 25°C	V_{CC} = 2.2 V	1.6	1.9	2.2	MHz
		V_{CC} = 3 V	1.69	2.0	2.29	
$f_{(DCO73)}$	R_{sel} = 7, DCO = 3, MOD = 0, DCOR = 0, T_A = 25°C	V_{CC} = 2.2 V	2.4	2.9	3.4	MHz
		V_{CC} = 3 V	2.7	3.2	3.65	
$f_{(DCO47)}$	R_{sel} = 4, DCO = 7, MOD = 0, DCOR = 0, T_A = 25°C	V_{CC} = 2.2 V/3 V	$f_{DCO40} \times 1.7$	$f_{DCO40} \times 2.1$	$f_{DCO40} \times 2.5$	MHz
$f_{(DCO77)}$	R_{sel} = 7, DCO = 7, MOD = 0, DCOR = 0, T_A = 25°C	V_{CC} = 2.2 V	4	4.5	4.9	MHz
		V_{CC} = 3 V	4.4	4.9	5.4	
$S_{(Rsel)}$	$S_R = f_{Rsel+1} / f_{Rsel}$	V_{CC} = 2.2 V/3 V	1.35	1.65	2	
$S_{(DCO)}$	$S_{DCO} = f_{DCO+1} / f_{DCO}$	V_{CC} = 2.2 V/3 V	1.07	1.12	1.16	
D_t	Temperature drift, R_{sel} = 4, DCO = 3, MOD = 0 (see Note 2)	V_{CC} = 2.2 V	–0.31	–0.36	–0.40	%/°C
		V_{CC} = 3 V	–0.33	–0.38	–0.43	
D_V	Drift with V_{CC} variation, R_{sel} = 4, DCO = 3, MOD = 0 (see Note 2)	V_{CC} = 2.2 V/3 V	0	5	10	%/V

NOTES: 1. The DCO frequency may not exceed the maximum system frequency defined by parameter processor frequency, $f_{(System)}$.
2. This parameter is not production tested.

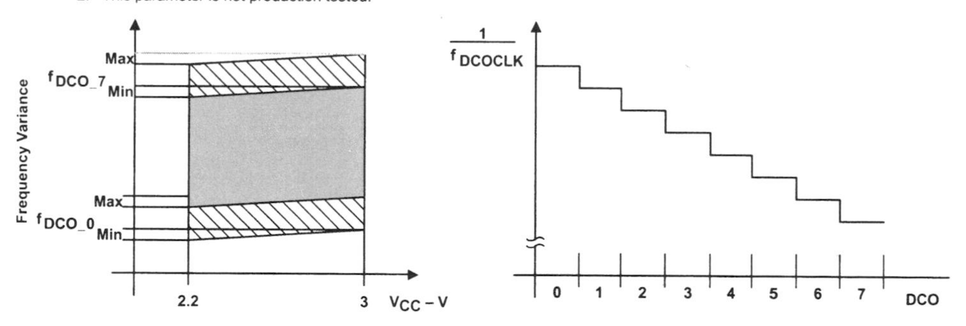

Figure 12. DCO Characteristics

MSP430x13x, MSP430x14x
MIXED SIGNAL MICROCONTROLLER

electrical characteristics over recommended operating free-air temperature (unless otherwise noted) (continued)

main DCO characteristics

- Individual devices have a minimum and maximum operation frequency. The specified parameters for fDCOx0 to fDCOx7 are valid for all devices.
- All ranges selected by Rsel(n) overlap with Rsel(n+1): Rsel0 overlaps with Rsel1, ... Rsel6 overlaps with Rsel7.
- DCO control bits DCO0, DCO1, and DCO2 have a step size as defined by parameter SDCO.
- Modulation control bits MOD0 to MOD4 select how often fDCO+1 is used within the period of 32 DCOCLK cycles. The frequency f(DCO) is used for the remaining cycles. The frequency is an average equal to f(DCO) × ($2^{MOD/32}$).

crystal oscillator, LFXT1 oscillator (see Note 1)

PARAMETER		TEST CONDITIONS	MIN	NOM	MAX	UNIT
X_{CIN}	Integrated input capacitance	XTS=0; LF oscillator selected V_{CC} = 2.2 V/3 V		12		pF
		XTS=1; XT1 oscillator selected V_{CC} = 2.2 V/3 V		2		
X_{COUT}	Integrated output capacitance	XTS=0; LF oscillator selected V_{CC} = 2.2 V/3 V		12		pF
		XTS=1; XT1 oscillator selected V_{CC} = 2.2 V/3 V		2		
X_{INL}	Input levels at XIN, XOUT	V_{CC} = 2.2 V/3 V	V_{SS}		0.2 × V_{CC}	V
X_{INH}		V_{CC} = 2.2 V/3 V	0.8 × V_{CC}		V_{CC}	V

NOTE 1: The oscillator needs capacitors at both terminals, with values specified by the crystal manufacturer.

crystal oscillator, XT2 oscillator (see Note 1)

PARAMETER		TEST CONDITIONS	MIN	NOM	MAX	UNIT
X_{CIN}	Integrated input capacitance	V_{CC} = 2.2 V/3 V		2		pF
X_{COUT}	Integrated output capacitance	V_{CC} = 2.2 V/3 V		2		pF
X_{INL}	Input levels at XIN, XOUT	V_{CC} = 2.2 V/3 V	V_{SS}		0.2 × V_{CC}	V
X_{INH}		V_{CC} = 2.2 V/3 V	0.8 × V_{CC}		V_{CC}	V

NOTE 1: The oscillator needs capacitors at both terminals, with values specified by the crystal manufacturer.

USART0, USART1 (see Note 1)

PARAMETER		TEST CONDITIONS	MIN	NOM	MAX	UNIT
$t_{(\tau)}$	USART0/1: deglitch time	V_{CC} = 2.2 V	200	430	800	ns
		V_{CC} = 3 V	150	280	500	

NOTE 1: The signal applied to the USART0/1 receive signal/terminal (URXD0/1) should meet the timing requirements of $t_{(t)}$ to ensure that the URXS flip-flop is set. The URXS flip-flop is set with negative pulses meeting the minimum-timing condition of $t_{(t)}$. The operating conditions to set the flag must be met independently from this timing constraint. The deglitch circuitry is active only on negative transitions on the URXD0/1 line.

MSP430x13x, MSP430x14x
MIXED SIGNAL MICROCONTROLLER

electrical characteristics over recommended operating free-air temperature (unless otherwise noted) (continued)

12-bit ADC, power supply and input range conditions (see Note 1)

PARAMETER		TEST CONDITIONS		MIN	NOM	MAX	UNIT
AV_{CC}	Analog supply voltage	AV_{CC} and DV_{CC} are connected together AV_{SS} and DV_{SS} are connected together $V_{(AVSS)} = V_{(DVSS)} = 0$ V		2.2		3.6	V
V_{REF+}	Positive built-in reference voltage output	REF2_5 V = 1 for 2.5 V built-in reference	3 V	2.4	2.5	2.6	V
		REF2_5 V = 0 for 1.5 V built-in reference $I_{V(REF+)} \le I_{(VREF+)max}$	2.2 V/3 V	1.44	1.5	1.56	
I_{VREF+}	Load current out of V_{REF+} terminal		2.2 V	0.01		–0.5	mA
			3 V			–1	
$I_{L(VREF)+}$ †	Load-current regulation V_{REF+} terminal	$I_{V(REF)+}$ = 500 µA +/– 100 µA Analog input voltage ~0.75 V; REF2_5 V = 0	2.2 V			±2	LSB
			3 V			±2	
		$I_{V(REF)+}$ = 500 µA ± 100 µA Analog input voltage ~1.25 V; REF2_5 V = 1	3 V			±2	LSB
$I_{DL(VREF)+}$ ‡	Load current regulation V_{REF+} terminal	$I_{V(REF)+}$ =100 µA → 900 µA, VCC=3 V, ax ~0.5 x V_{REF+} Error of conversion result ≤ 1 LSB	C_{VREF+}=5 µF			20	ns
V_{eREF+}	Positive external reference voltage input	$V_{eREF+} > V_{REF-}/V_{eREF-}$ (see Note 2)		1.4		V_{AVCC}	V
V_{REF-}/V_{eREF-}	Negative external reference voltage input	$V_{eREF+} > V_{REF-}/V_{eREF-}$ (see Note 3)		0		1.2	V
(V_{eREF+} – V_{REF-}/V_{eREF-})	Differential external reference voltage input	$V_{eREF+} > V_{REF-}/V_{eREF-}$ (see Note 4)		1.4		V_{AVCC}	V
$V_{(P6.x/Ax)}$	Analog input voltage range (see Note 5)	All P6.0/A0 to P6.7/A7 terminals. Analog inputs selected in ADC12MCTLx register and P6Sel.x=1 $0 \le x \le 7$; $V_{(AVSS)} \le V_{P6.x/Ax} \le V_{(AVCC)}$		0		V_{AVCC}	V
I_{ADC12}	Operating supply current into AV_{CC} terminal (see Note 6)	$f_{ADC12CLK}$ = 5.0 MHz ADC12ON = 1, REFON = 0 SHT0=0, SHT1=0, ADC12DIV=0	2.2 V		0.65	1.3	mA
			3 V		0.8	1.6	
I_{REF+}	Operating supply current into AV_{CC} terminal (see Note 7)	$f_{ADC12CLK}$ = 5.0 MHz ADC12ON = 0, REFON = 1, REF2_5V = 1	3 V		0.5	0.8	mA
I_{REF+}	Operating supply current (see Note 7)	$f_{ADC12CLK}$ = 5.0 MHz ADC12ON = 0, REFON = 1, REF2_5V = 0	2.2 V		0.5	0.8	mA
			3 V		0.5	0.8	

† Not production tested, limits characterized
‡ Not production tested, limits verified by design

NOTES:
1. The leakage current is defined in the leakage current table with P6.x/Ax parameter.
2. The accuracy limits the minimum positive external reference voltage. Lower reference voltage levels may be applied with reduced accuracy requirements.
3. The accuracy limits the maximum negative external reference voltage. Higher reference voltage levels may be applied with reduced accuracy requirements.
4. The accuracy limits minimum external differential reference voltage. Lower differential reference voltage levels may be applied with reduced accuracy requirements.
5. The analog input voltage range must be within the selected reference voltage range V_{R+} to V_{R-} for valid conversion results.
6. The internal reference supply current is not included in current consumption parameter I_{ADC12}.
7. The internal reference current is supplied via terminal AV_{CC}. Consumption is independent of the ADC12ON control bit, unless a conversion is active. The REFON bit enables to settle the built-in reference before starting an A/D conversion.

MSP430x13x, MSP430x14x
MIXED SIGNAL MICROCONTROLLER

electrical characteristics over recommended operating free-air temperature (unless otherwise noted) (continued)

12-bit ADC, built-in reference (see Note 1)

PARAMETER		TEST CONDITIONS		MIN	NOM	MAX	UNIT
I_{VeREF+}	Static input current (see Note 2)	$0V \leq V_{eREF+} \leq V_{AVCC}$	2.2 V/3 V			±1	µA
$I_{VREF-/VeREF-}$	Static input current (see Note 2)	$0V \leq V_{eREF-} \leq V_{AVCC}$	2.2 V/3 V			±1	µA
C_{VREF+}	Capacitance at pin V_{REF+} (see Note 3)	REFON =1, $0\ mA \leq I_{VREF+} \leq I_{V(REF)+(max)}$	2.2 V/3 V	5	10		µF
C_i ‡	Input capacitance (see Note 4)	Only one terminal can be selected at one time, P6.x/Ax	2.2 V			40	pF
Z_i‡	Input MUX ON resistance (see Note 4)	$0V \leq V_{Ax} \leq V_{AVCC}$	3 V			2000	Ω
T_{REF+}†	Temperature coefficient of built-in reference	$I_{V(REF)+}$ is a constant in the range of $0\ mA \leq I_{V(REF)+} \leq 1\ mA$	2.2 V/3 V			±100	ppm/°C

† Not production tested, limits characterized

‡ Not production tested, limits verified by design

NOTES: 1. The voltage source on V_{eREF+} and V_{REF-}/V_{eREF-}) needs to have low dynamic impedance for 12-bit accuracy to allow the charge to settle for this accuracy.

2. The external reference is used during conversion to charge and discharge the capacitance array. The dynamic impedance should follow the recommendations on analog-source impedance to allow the charge to settle for 12-bit accuracy.

3. The internal buffer operational amplifier and the accuracy specifications require an external capacitor.

4. The input capacitance is also the dynamic load for an external reference during conversion. The dynamic impedance of the reference supply should follow the recommendations on analog-source impedance to allow the charge to settle for 12-bit accuracy. All INL and DNL tests uses two capacitors between pins V_{REF+} and AV_{SS} and V_{REF-}/V_{eREF-} and AV_{SS}: 10µF tantalum and 100 nF ceramic.

MSP430x13x, MSP430x14x
MIXED SIGNAL MICROCONTROLLER

electrical characteristics over recommended operating free-air temperature (unless otherwise noted) (continued)

12-bit ADC, timing parameters

PARAMETER		TEST CONDITIONS		MIN	NOM	MAX	UNIT
$t_{REF(ON)}$†	Settle time of internal reference voltage (see Figure 13 and Note 1)	$I_{V(REF)+}$ = 0.5 mA, $C_{V(REF)+}$ = 10 μF, V_{REF+} = 1.5 V, V_{AVCC} = 2.2 V				17	ms
f(ADC12CLK)		Error of conversion result ≤ ±2 LSB	2.2V/ 3V		5		MHz
f(ADC12OSC)		ADC12DIV=0 [f(ADC12CLK) =f(ADC12OSC)]	2.2 V/ 3V	3.7		6.3	MHz
$t_{CONVERT}$	Conversion time	$AV_{CC(min)} \le V_{AVCC} \le AV_{CC(max)}$, $C_{VREF+} \ge 5$ μF, Internal oscillator, f_{OSC} = 3.7 MHz to 6.3 MHz	2.2 V/ 3 V	2.06		3.51	μs
	Conversion time	$AV_{CC(min)} \le V_{AVCC} \le AV_{CC(max)}$, External $f_{ADC12CLK}$ from ACLK or MCLK or SMCLK: ADC12SSEL ≠ 0			13×ADC12DIV× 1/$f_{ADC12CLK}$		μs
$t_{ADC12ON}$‡	Settle time of the ADC	$AV_{CC(min)} \le V_{AVCC} \le AV_{CC(max)}$ (see Note 2)				100	ns
t_{Sample}‡	Sampling time	$V_{AVCC(min)} \le V_{AVCC} \le V_{AVCC(max)}$, $R_{i(source)}$ = 400 Ω, Z_i = 1000 Ω, C_i = 30 pF, $\tau = [R_{i(source)} \times + Z_i] \times C_i$;(see Note 3)	3 V	1220			ns
			2.2 V	1400			ns

† Not production tested, limits characterized

‡ Not production tested, limits verified by design

NOTES: 1. The condition is that the error in a conversion started after $t_{REF(ON)}$ is less than ±0.5 LSB. The settling time depends on the external capacitive load.

2. The condition is that the error in a conversion started after $t_{ADC12ON}$ is less than ±0.5 LSB. The reference and input signal are already settled.

3. Ten Tau (τ) are needed to get an error of less than ±0.5 LSB. t_{Sample} = 10 x (Ri + Zi) x Ci+ 800 ns

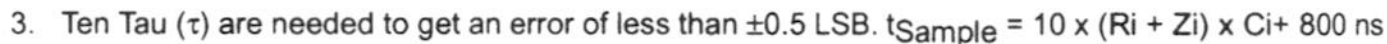

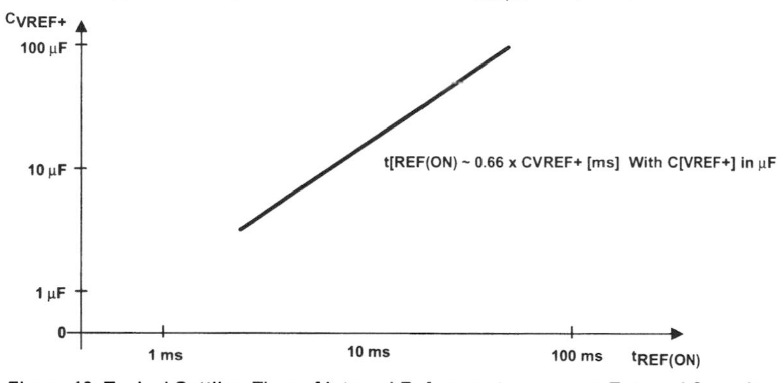

Figure 13. Typical Settling Time of Internal Reference $t_{REF(ON)}$ vs External Capacitor on V_{REF+}

MSP430x13x, MSP430x14x MIXED SIGNAL MICROCONTROLLER

electrical characteristics over recommended operating free-air temperature (unless otherwise noted) (continued)

12-bit ADC, linearity parameters

PARAMETER		TEST CONDITIONS		MIN	NOM	MAX	UNIT
$E_{(I)}$	Integral linearity error	1.4 V ≤ (VeREF+ – VREF–/VeREF–) min ≤ 1.6 V	2.2 V/3 V			±2	LSB
		1.6 V < (VeREF+ – VREF–/VeREF–) min ≤ [V(AVCC)]				±1.7	
E_D	Differential linearity error	$(V_{eREF+} - V_{REF-}/V_{eREF-})_{min} \le (V_{eREF+} - V_{REF-}/V_{eREF-})$, C(VREF+) = 10 µF (tantalum) and 100 nF (ceramic)	2.2 V/3 V			±1	LSB
E_O	Offset error	$(V_{eREF+} - V_{REF-}/V_{eREF-})_{min} \le (V_{eREF+} - V_{REF-}/V_{eREF-})$, Internal impedance of source R_i < 100 Ω, C(VREF+) = 10 µF (tantalum) and 100 nF (ceramic)	2.2 V/3 V		±2	±4	LSB
E_G	Gain error	$(V_{eREF+} - V_{REF-}/V_{eREF-})_{min} \le (V_{eREF+} - V_{REF-}/V_{eREF-})$, C(VREF+) = 10 µF (tantalum) and 100 nF (ceramic)	2.2 V/3 V		±1.1	±2	LSB
E_T	Total unadjusted error	$(V_{eREF+} - V_{REF-}/V_{eREF-})_{min} \le (V_{eREF+} - V_{REF-}/V_{eREF-})$, C(VREF+) = 10 µF (tantalum) and 100 nF (ceramic)	2.2 V/3 V		±2	±5	LSB

MSP430x13x, MSP430x14x
MIXED SIGNAL MICROCONTROLLER

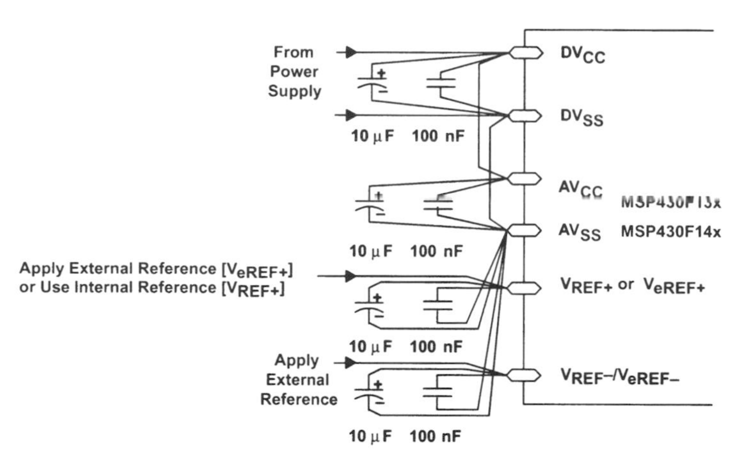

Figure 14. Supply Voltage and Reference Voltage Design $V_{(REF-)}/V_{(eREF-)}$ External Supply

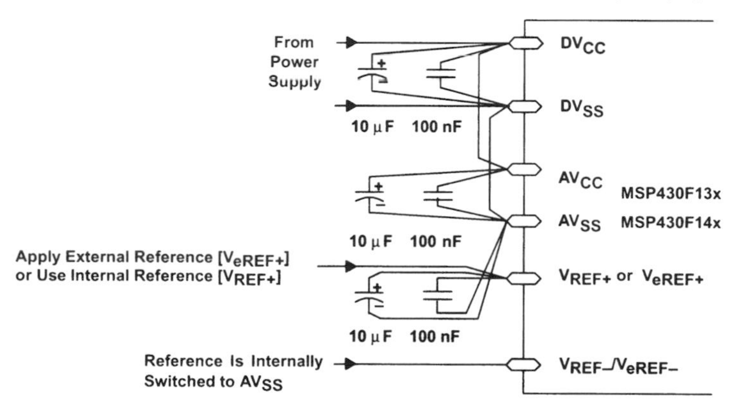

Figure 15. Supply Voltage and Reference Voltage Design $V_{REF-}/V_{eREF-} = AV_{SS}$, Internally Connected

MSP430x13x, MSP430x14x
MIXED SIGNAL MICROCONTROLLER

electrical characteristics over recommended operating free-air temperature (unless otherwise noted) (continued)

12-bit ADC, temperature sensor and built-in Vmid

PARAMETER		TEST CONDITIONS		MIN	NOM	MAX	UNIT
I_{SENSOR}	Operating supply current into AV_{CC} terminal (see Note 1)	V_{REFON} = 0, INCH = 0Ah, ADC12ON=NA, T_A = 25°C	2.2 V		40	120	µA
			3 V		60	160	
V_{SENSOR}†		ADC12ON = 1, INCH = 0Ah, T_A = 0°C	2.2 V		986	986±5%	mV
			3 V		986	986±5%	
TC_{SENSOR}†		ADC12ON = 1, INCH = 0Ah	2.2 V		3.55	3.55±3%	mV/°C
			3 V		3.55	3.55±3%	
$t_{SENSOR(sample)}$†	Sample time required if channel 10 is selected (see Note 2)	ADC12ON = 1, INCH = 0Ah, Error of conversion result ≤ 1 LSB	2.2 V	30			µs
			3 V	30			
I_{VMID}	Current into divider at channel 11	ADC12ON = 1, INCH = 0Bh, (see Note 3)	2.2 V			NA	µA
			3 V			NA	
V_{MID}	AV_{CC} divider at channel 11	ADC12ON = 1, INCH = 0Bh, V_{MID} is ~0.5 x V_{AVCC}	2.2 V		1.1	1.1±0.04	V
			3 V		1.5	1.50±0.04	
$t_{ON(VMID)}$	On-time if channel 11 is selected (see Note 4)	ADC12ON = 1, INCH = 0Bh, Error of conversion result ≤ 1 LSB	2.2 V			NA	ns
			3 V			NA	

† Not production tested, limits characterized
‡ Not production tested, limits verified by design

NOTES: 1. The sensor current I_{SENSOR} is consumed if (ADC12ON = 1 and V_{REFON}=1), or (ADC12ON=1 AND INCH=0Ah and sample signal is high). Therefore it includes the constant current through the sensor and the reference.
2. The typical equivalent impedance of the sensor is 51 kΩ. The sample time needed is the sensor-on time $t_{SENSOR(ON)}$
3. No additional current is needed. The V_{MID} is used during sampling.
4. The on-time $t_{ON(VMID)}$ is identical to sampling time t_{Sample}; no additional on time is needed.

JTAG, program memory and fuse

PARAMETER		TEST CONDITIONS	V_{CC}	MIN	NOM	MAX	UNIT
$f_{(TCK)}$	JTAG/Test (see Note 4)	TCK frequency	2.2 V	DC		5	MHz
			3 V	DC		10	
		Pullup resistors on TMS, TCK, TDI (see Note 1)	2.2 V/ 3V	25	60	90	kΩ
$V_{CC(FB)}$	JTAG/fuse (see Note 2)	Supply voltage during fuse-blow condition, $T_{(A)}$ = 25°C		2.5			V
V_{FB}		Fuse-blow voltage, F versions (see Note 3)		6.0		7.0	V
I_{FB}		Supply current on TDI with fuse blown				100	mA
		Time to blow the fuse				1	ms
$I_{(DD-PGM)}$	F-versions only (see Note 4)	Current from DV_{CC} when programming is active	2.7 V/3.6 V		3	5	mA
$I_{(DD-Erase)}$		Current from DV_{CC} when erase is active	2.7 V/3.6 V		3	5	mA
$t_{(retention)}$	F-versions only	Write/erase cycles		10^4	10^5		cycles
		Data retention T_J = 25°C		100			years

NOTES: 1. TMS, TDI, and TCK pull-up resistors are implemented in all F versions.
2. Once the fuse is blown, no further access to the MSP430 JTAG/test feature is possible. The JTAG block is switched to bypass mode.
3. The supply voltage to blow the fuse is applied to the TDI pin.
4. $f_{(TCK)}$ may be restricted to meet the timing requirements of the module selected. Duration of the program/erase cycle is determined by $f_{(FTG)}$ applied to the flash timing controller. It can be calculated as follows:
$t_{(word\ write)} = 35 \times 1/f_{(FTG)}$
$t_{(block\ write,\ byte\ 0)} = 30 \times 1/f_{(FTG)}$
$t_{(block\ write,\ bytes\ 1–63)} = 20 \times 1/f_{(FTG)}$
$t_{(block\ write\ end\ sequence)} = 6 \times 1/f_{(FTG)}$
$t_{(mass\ erase)} = 5297 \times 1/f_{(FTG)}$
$t_{(segment\ erase)} = 4819 \times 1/f_{(FTG)}$

MSP430x13x, MSP430x14x
MIXED SIGNAL MICROCONTROLLER

input/output schematic

port P1, P1.0 to P1.7, input/output with Schmitt-trigger

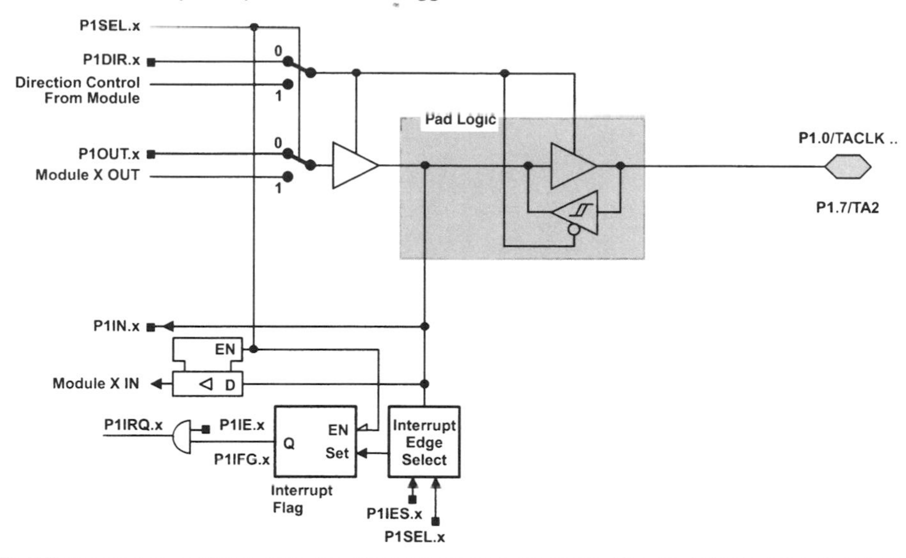

PnSel.x	PnDIR.x	Dir. CONTROL FROM MODULE	PnOUT.x	MODULE X OUT	PnIN.x	MODULE X IN	PnIE.x	PnIFG.x	PnIES.x
P1Sel.0	P1DIR.0	P1DIR.0	P1OUT.0	DV_{SS}	P1IN.0	TACLK†	P1IE.0	P1IFG.0	P1IES.0
P1Sel.1	P1DIR.1	P1DIR.1	P1OUT.1	Out0 signal†	P1IN.1	CCI0A†	P1IE.1	P1IFG.1	P1IES.1
P1Sel.2	P1DIR.2	P1DIR.2	P1OUT.2	Out1 signal†	P1IN.2	CCI1A†	P1IE.2	P1IFG.2	P1IES.2
P1Sel.3	P1DIR.3	P1DIR.3	P1OUT.3	Out2 signal†	P1IN.3	CCI2A†	P1IE.3	P1IFG.3	P1IES.3
P1Sel.4	P1DIR.4	P1DIR.4	P1OUT.4	SMCLK	P1IN.4	unused	P1IE.4	P1IFG.4	P1IES.4
P1Sel.5	P1DIR.5	P1DIR.5	P1OUT.5	Out0 signal†	P1IN.5	unused	P1IE.5	P1IFG.5	P1IES.5
P1Sel.6	P1DIR.6	P1DIR.6	P1OUT.6	Out1 signal†	P1IN.6	unused	P1IE.6	P1IFG.6	P1IES.6
P1Sel.7	P1DIR.7	P1DIR.7	P1OUT.7	Out2 signal†	P1IN.7	unused	P1IE.7	P1IFG.7	P1IES.7

† Signal from or to Timer_A

MSP430x13x, MSP430x14x
MIXED SIGNAL MICROCONTROLLER

input/output schematic (continued)

port P2, P2.0 to P2.2, P2.6, and P2.7 input/output with Schmitt-trigger

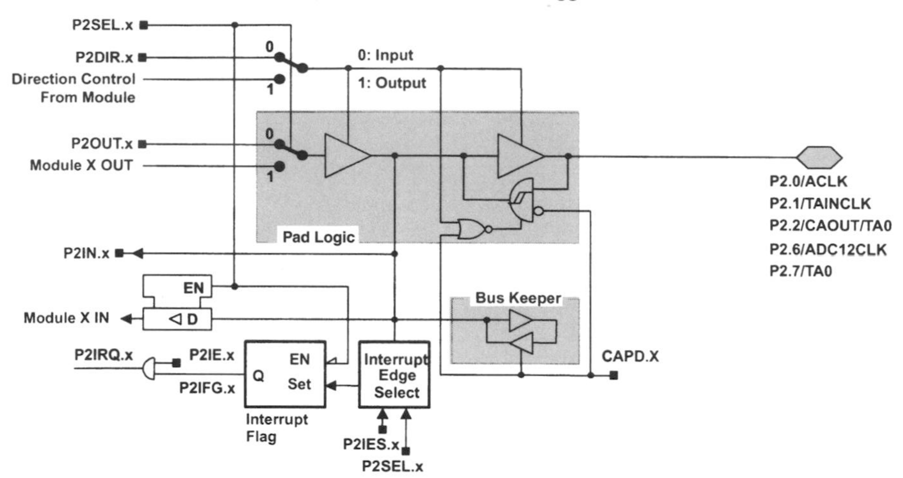

x: Bit Identifier 0 to 2, 6, and 7 for Port P2

PnSel.x	PnDIR.x	Dir. CONTROL FROM MODULE	PnOUT.x	MODULE X OUT	PnIN.x	MODULE X IN	PnIE.x	PnIFG.x	PnIES.x
P2Sel.0	P2DIR.0	P2DIR.0	P2OUT.0	ACLK	P2IN.0	unused	P2IE.0	P2IFG.0	P2IES.0
P2Sel.1	P2DIR.1	P2DIR.1	P2OUT.1	DV_{SS}	P2IN.1	INCLK‡	P2IE.1	P2IFG.1	P2IES.1
P2Sel.2	P2DIR.2	P2DIR.2	P2OUT.2	CAOUT†	P2IN.2	CCI0B‡	P2IE.2	P2IFG.2	P2IES.2
P2Sel.6	P2DIR.6	P2DIR.6	P2OUT.6	ADC12CLK¶	P2IN.6	unused	P2IE.6	P2IFG.6	P2IES.6
P2Sel.7	P2DIR.7	P2DIR.7	P2OUT.7	Out0 signal§	P2IN.7	unused	P2IE.7	P2IFG.7	P2IES.7

† Signal from Comparator_A
‡ Signal to Timer_A
§ Signal from Timer_A
¶ ADC12CLK signal is output of the 12-bit ADC module

input/output schematic (continued)

port P2, P2.3 to P2.4, input/output with Schmitt-trigger

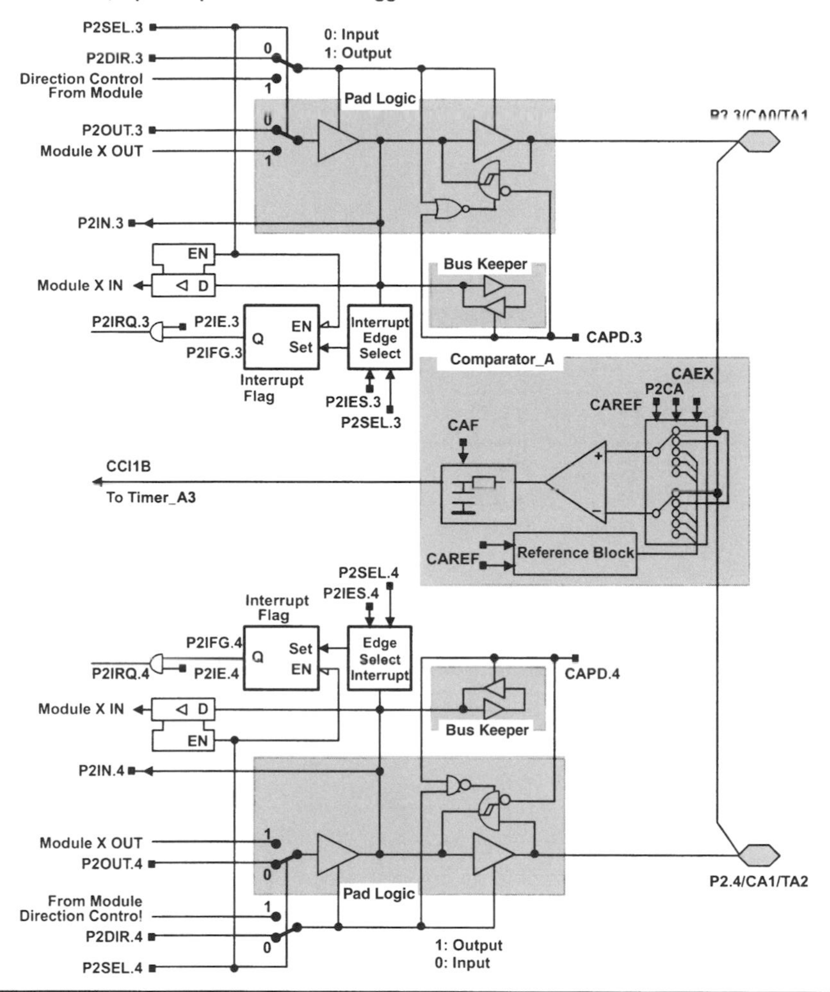

PnSel.x	PnDIR.x	DIRECTION CONTROL FROM MODULE	PnOUT.x	MODULE X OUT	PnIN.x	MODULE X IN	PnIE.x	PnIFG.x	PnIES.x
P2Sel.3	P2DIR.3	P2DIR.3	P2OUT.3	Out1 signal†	P2IN.3	unused	P2IE.3	P2IFG.3	P2IES.3
P2Sel.4	P2DIR.4	P2DIR.4	P2OUT.4	Out2 signal†	P2IN.4	unused	P2IE.4	P2IFG.4	P2IES.4

† Signal from Timer_A

MSP430x13x, MSP430x14x
MIXED SIGNAL MICROCONTROLLER

input/output schematic (continued)

port P2, P2.5, input/output with Schmitt-trigger and R_{osc} function for the basic clock module

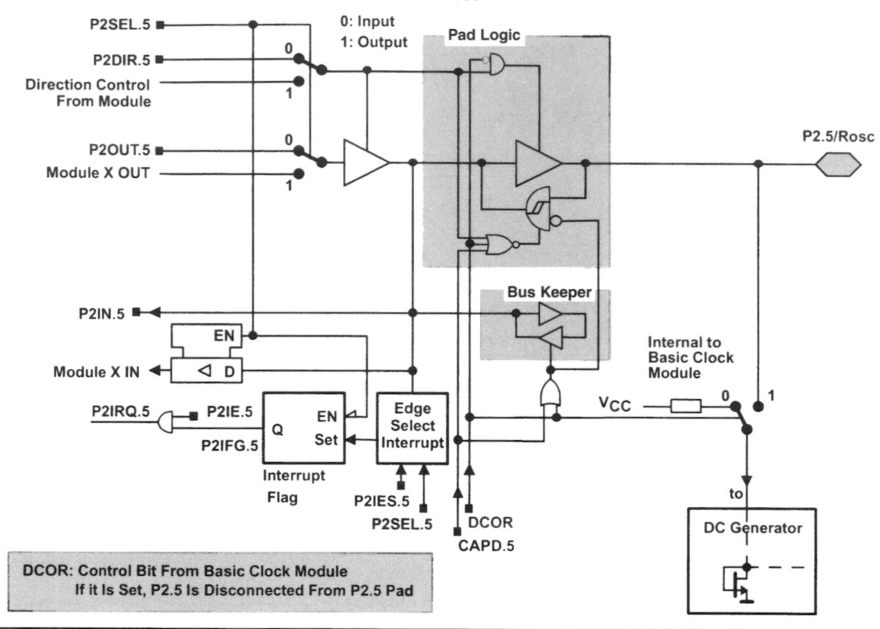

PnSel.x	PnDIR.x	DIRECTION CONTROL FROM MODULE	PnOUT.x	MODULE X OUT	PnIN.x	MODULE X IN	PnIE.x	PnIFG.x	PnIES.x
P2Sel.5	P2DIR.5	P2DIR.5	P2OUT.5	DV_{SS}	P2IN.5	unused	P2IE.5	P2IFG.5	P2IES.5

input/output schematic (continued)

port P3, P3.0 and P3.4 to P3.7, input/output with Schmitt-trigger

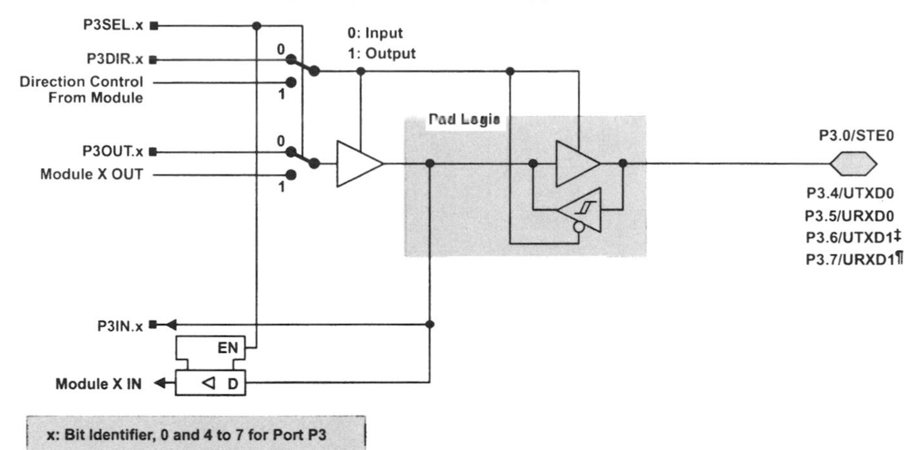

x: Bit Identifier, 0 and 4 to 7 for Port P3

PnSel.x	PnDIR.x	DIRECTION CONTROL FROM MODULE	PnOUT.x	MODULE X OUT	PnIN.x	MODULE X IN
P3Sel.0	P3DIR.0	DV_{SS}	P3OUT.0	DV_{SS}	P3IN.0	STE0
P3Sel.4	P3DIR.4	DV_{CC}	P3OUT.4	UTXD0†	P3IN.4	Unused
P3Sel.5	P3DIR.5	DV_{SS}	P3OUT.5	DV_{SS}	P3IN.5	URXD0§
P3Sel.6	P3DIR.6	DV_{CC}	P3OUT.6	UTXD1‡	P3IN.6	Unused
P3Sel.7	P3DIR.7	DV_{SS}	P3OUT.7	DV_{SS}	P3IN.7	URXD1¶

† Output from USART0 module
‡ Output from USART1 module in x14x configuration, DV_{SS} in x13x configuration
§ Input to USART0 module
¶ Input to USART1 module in x14x configuration, unused in x13x configuration

port P3, P3.1, input/output with Schmitt-trigger

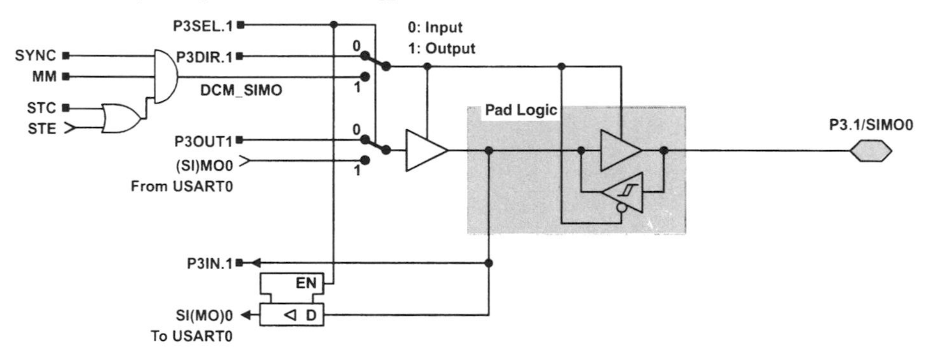

MSP430x13x, MSP430x14x
MIXED SIGNAL MICROCONTROLLER

input/output schematic (continued)

port P3, P3.2, input/output with Schmitt-trigger

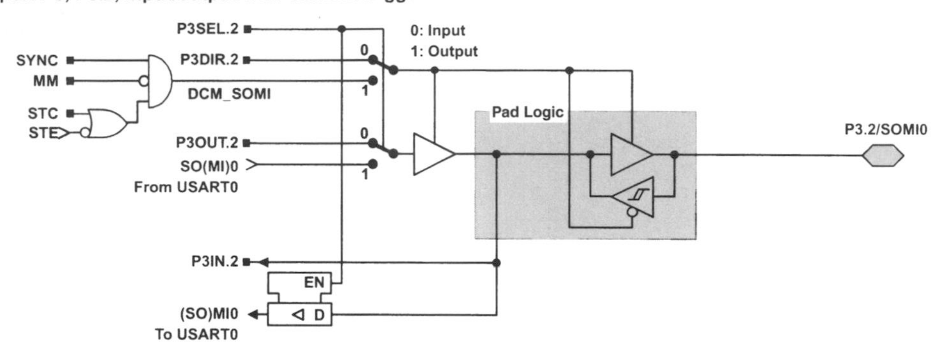

port P3, P3.3, input/output with Schmitt-trigger

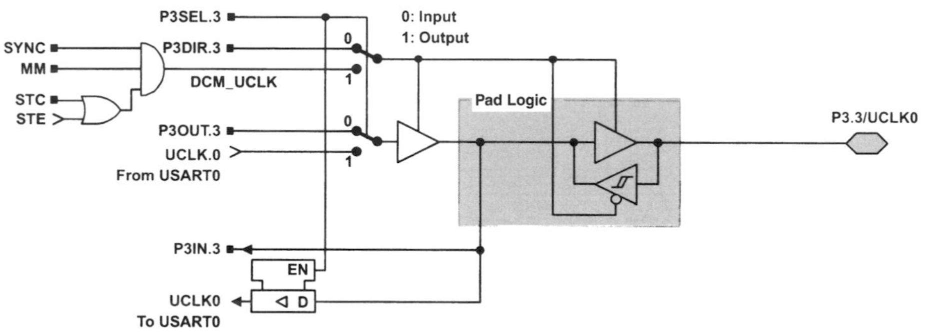

NOTE: UART mode: The UART clock can only be an input. If UART mode and UART function are selected, the P3.3/UCLK0 is always an input.

SPI, slave mode: The clock applied to UCLK0 is used to shift data in and out.

SPI, master mode: The clock to shift data in and out is supplied to connected devices on pin P3.3/UCLK0 (in slave mode).

MSP430x13x, MSP430x14x
MIXED SIGNAL MICROCONTROLLER

input/output schematic (continued)

port P4, P4.0 to P4.6, input/output with Schmitt-trigger

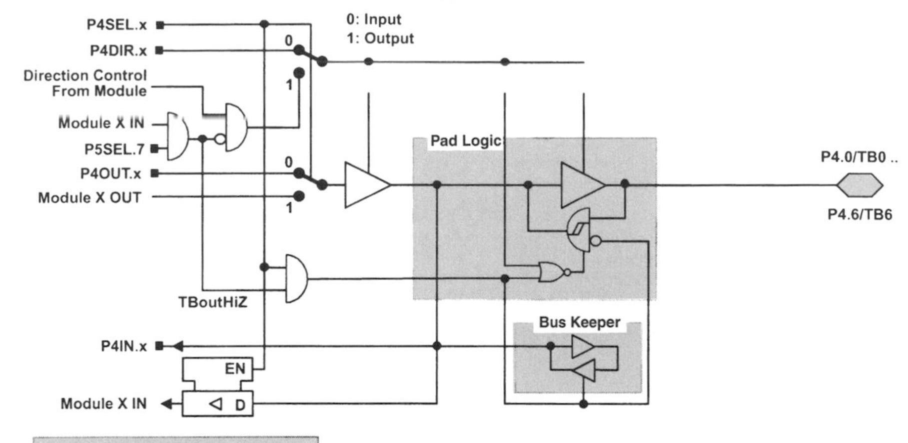

x: bit identifier, 0 to 6 for Port P4

PnSel.x	PnDIR.x	DIRECTION CONTROL FROM MODULE	PnOUT.x	MODULE X OUT	PnIN.x	MODULE X IN
P4Sel.0	P4DIR.0	P4DIR.0	P4OUT.0	Out0 signal†	P4IN.0	CCI0A / CCI0B‡
P4Sel.1	P4DIR.1	P4DIR.1	P4OUT.1	Out1 signal†	P4IN.1	CCI1A / CCI1B‡
P4Sel.2	P4DIR.2	P4DIR.2	P4OUT.2	Out2 signal†	P4IN.2	CCI2A / CCI2B‡
P4Sel.3	P4DIR.3	P4DIR.3	P4OUT.3	Out3 signal†	P4IN.3	CCI3A / CCI3B‡
P4Sel.4	P4DIR.4	P4DIR.4	P4OUT.4	Out4 signal†	P4IN.4	CCI4A / CCI4B‡
P4Sel.5	P4DIR.5	P4DIR.5	P4OUT.5	Out5 signal†	P4IN.5	CCI5A / CCI5B‡
P4Sel.6	P4DIR.6	P4DIR.6	P4OUT.6	Out6 signal†	P4IN.6	CCI6A / CCI6B‡

† Signal from Timer_B
‡ Signal to Timer_B

MSP430x13x, MSP430x14x
MIXED SIGNAL MICROCONTROLLER

input/output schematic (continued)

port P4, P4.7, input/output with Schmitt-trigger

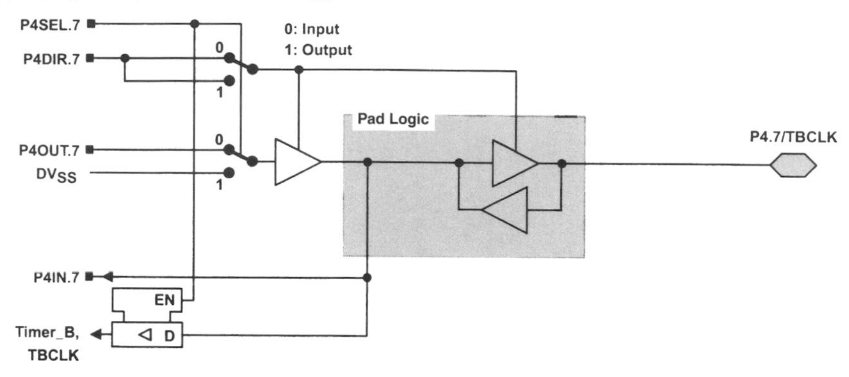

port P5, P5.0 and P5.4 to P5.7, input/output with Schmitt-trigger

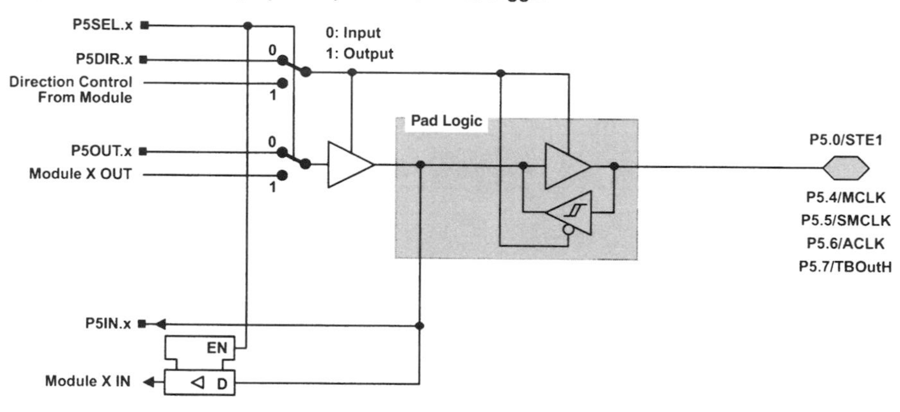

PnSel.x	PnDIR.x	Dir. CONTROL FROM MODULE	PnOUT.x	MODULE X OUT	PnIN.x	MODULE X IN
P5Sel.0	P5DIR.0	DV_{SS}	P5OUT.0	DV_{SS}	P5IN.0	STE.1
P5Sel.4	P5DIR.4	DV_{CC}	P5OUT.4	MCLK	P5IN.4	unused
P5Sel.5	P5DIR.5	DV_{CC}	P5OUT.5	SMCLK	P5IN.5	unused
P5Sel.6	P5DIR.6	DV_{CC}	P5OUT.6	ACLK	P5IN.6	unused
P5Sel.7	P5DIR.7	DV_{SS}	P5OUT.7	DV_{SS}	P5IN.7	TBoutHiZ

NOTE: TBoutHiZ signal is used by port module P4, pins P4.0 to P4.6. The function of TboutHiZ is mainly useful when used with Timer_B7.

input/output schematic (continued)

port P5, P5.1, input/output with Schmitt-trigger

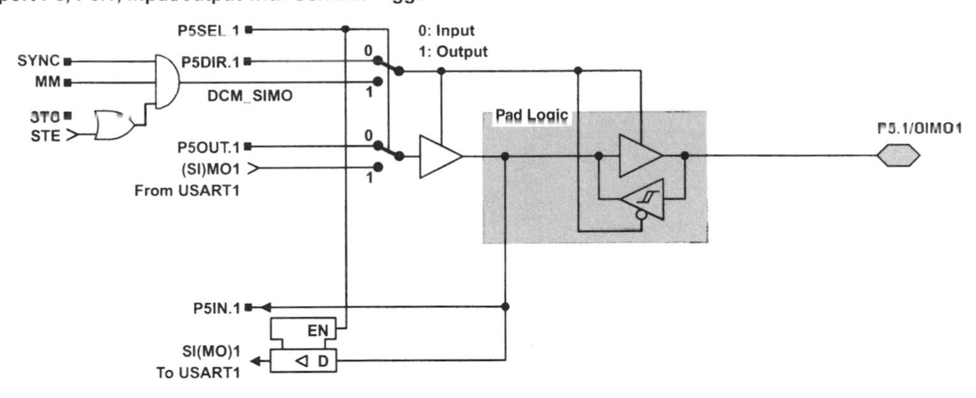

port P5, P5.2, input/output with Schmitt-trigger

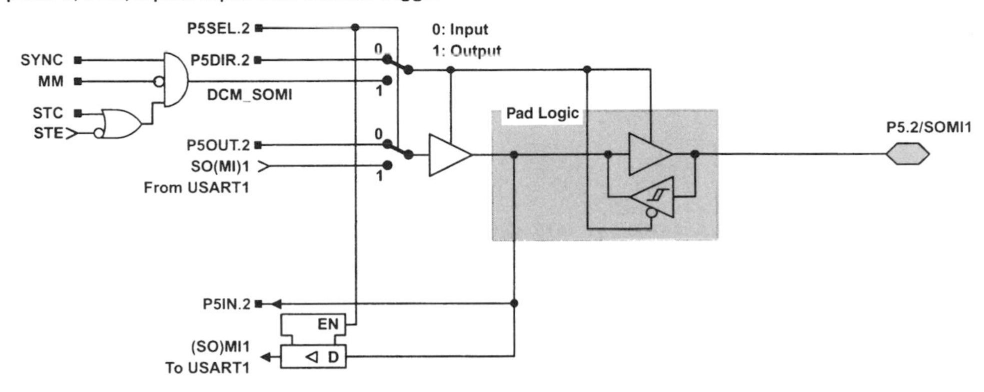

MSP430x13x, MSP430x14x
MIXED SIGNAL MICROCONTROLLER

input/output schematic (continued)

port P5, P5.3, input/output with Schmitt-trigger

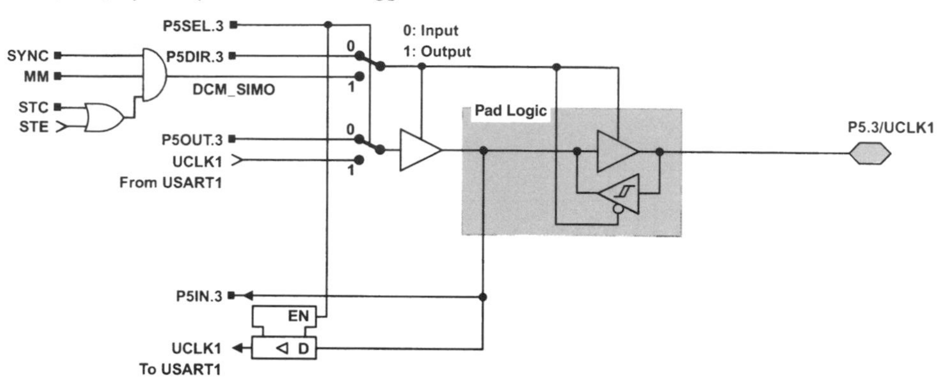

NOTE:		
	UART mode:	The UART clock can only be an input. If UART mode and UART function are selected, the P5.3/UCLK1 direction is always input.
	SPI, slave mode:	The clock applied to UCLK1 is used to shift data in and out.
	SPI, master mode:	The clock to shift data in and out is supplied to connected devices on pin P5.3/UCLK1 (in slave mode).

MSP430x13x, MSP430x14x
MIXED SIGNAL MICROCONTROLLER

input/output schematic (continued)

port P6, P6.0 to P6.7, input/output with Schmitt-trigger

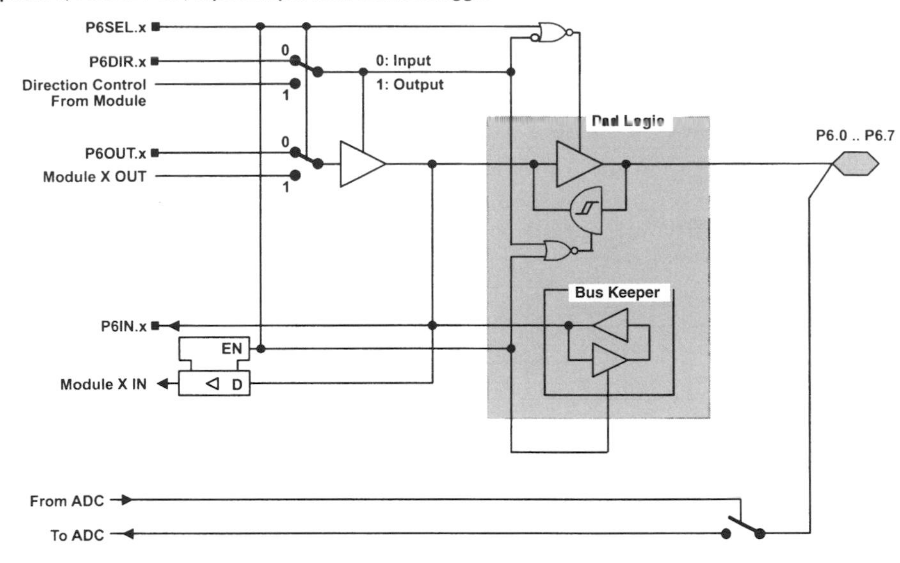

x: Bit Identifier, 0 to 7 for Port P6

NOTE: Analog signals applied to digital gates can cause current flow from the positive to the negative terminal. The throughput current flows if the analog signal is in the range of transitions 0→1 or 1←0. The value of the throughput current depends on the driving capability of the gate. For MSP430, it is approximately 100 µA.
Use P6SEL.x=1 to prevent throughput current. P6SEL.x should be set, even if the signal at the pin is not being used by the ADC12.

PnSel.x	PnDIR.x	DIR. CONTROL FROM MODULE	PnOUT.x	MODULE X OUT	PnIN.x	MODULE X IN
P6Sel.0	P6DIR.0	P6DIR.0	P6OUT.0	DV_{SS}	P6IN.0	unused
P6Sel.1	P6DIR.1	P6DIR.1	P6OUT.1	DV_{SS}	P6IN.1	unused
P6Sel.2	P6DIR.2	P6DIR.2	P6OUT.2	DV_{SS}	P6IN.2	unused
P6Sel.3	P6DIR.3	P6DIR.3	P6OUT.3	DV_{SS}	P6IN.3	unused
P6Sel.4	P6DIR.4	P6DIR.4	P6OUT.4	DV_{SS}	P6IN.4	unused
P6Sel.5	P6DIR.5	P6DIR.5	P6OUT.5	DV_{SS}	P6IN.5	unused
P6Sel.6	P6DIR.6	P6DIR.6	P6OUT.6	DV_{SS}	P6IN.6	unused
P6Sel.7	P6DIR.7	P6DIR.7	P6OUT.7	DV_{SS}	P6IN.7	unused

NOTE: The signal at pins P6.x/Ax is used by the 12-bit ADC module.

MSP430x13x, MSP430x14x
MIXED SIGNAL MICROCONTROLLER

input/output schematic (continued)

JTAG pins TMS, TCK, TDI, TDO/TDI, input/output with Schmitt-trigger

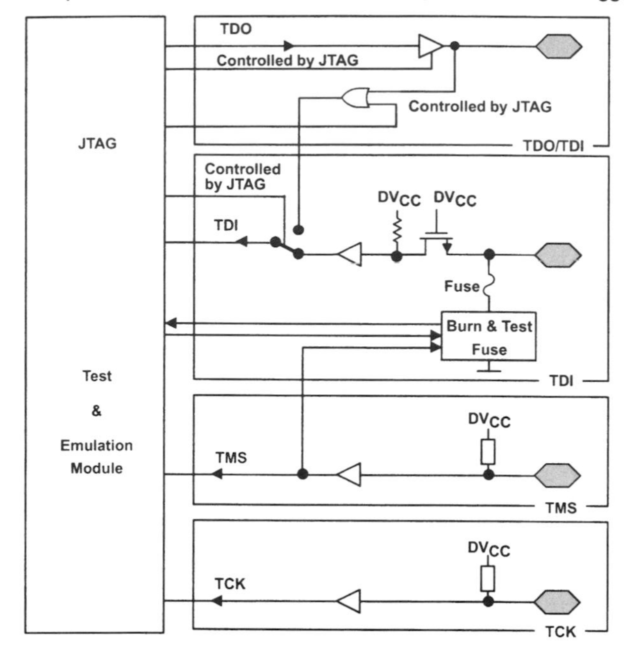

During Programming Activity and During Blowing of the Fuse, Pin TDO/TDI Is Used to Apply the Test Input Data for JTAG Circuitry

JTAG fuse check mode

MSP430 devices that have the fuse on the TDI terminal have a fuse check mode that tests the continuity of the fuse the first time the JTAG port is accessed after a power-on reset (POR). When activated, a fuse check current, I_{TF}, of 1 mA at 3 V, 2.5 mA at 5 V can flow from the TDI pin to ground if the fuse is not burned. Care must be taken to avoid accidentally activating the fuse check mode and increasing overall system power consumption.

Activation of the fuse check mode occurs with the first negative edge on the TMS pin after power up or if the TMS is being held low during power up. The second positive edge on the TMS pin deactivates the fuse check mode. After deactivation, the fuse check mode remains inactive until another POR occurs. After each POR the fuse check mode has the potential to be activated.

The fuse check current will only flow when the fuse check mode is active and the TMS pin is in a low state (see Figure 16). Therefore, the additional current flow can be prevented by holding the TMS pin high (default condition).

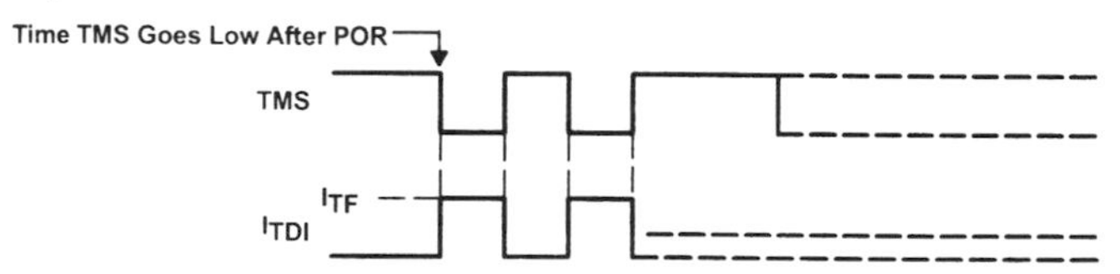

Figure 16. Fuse Check Mode Current, MSP430F13x, MSP430F14x

MSP430x13x, MSP430x14x
MIXED SIGNAL MICROCONTROLLER

MECHANICAL DATA

PM (S-PQFP-G64) PLASTIC QUAD FLATPACK

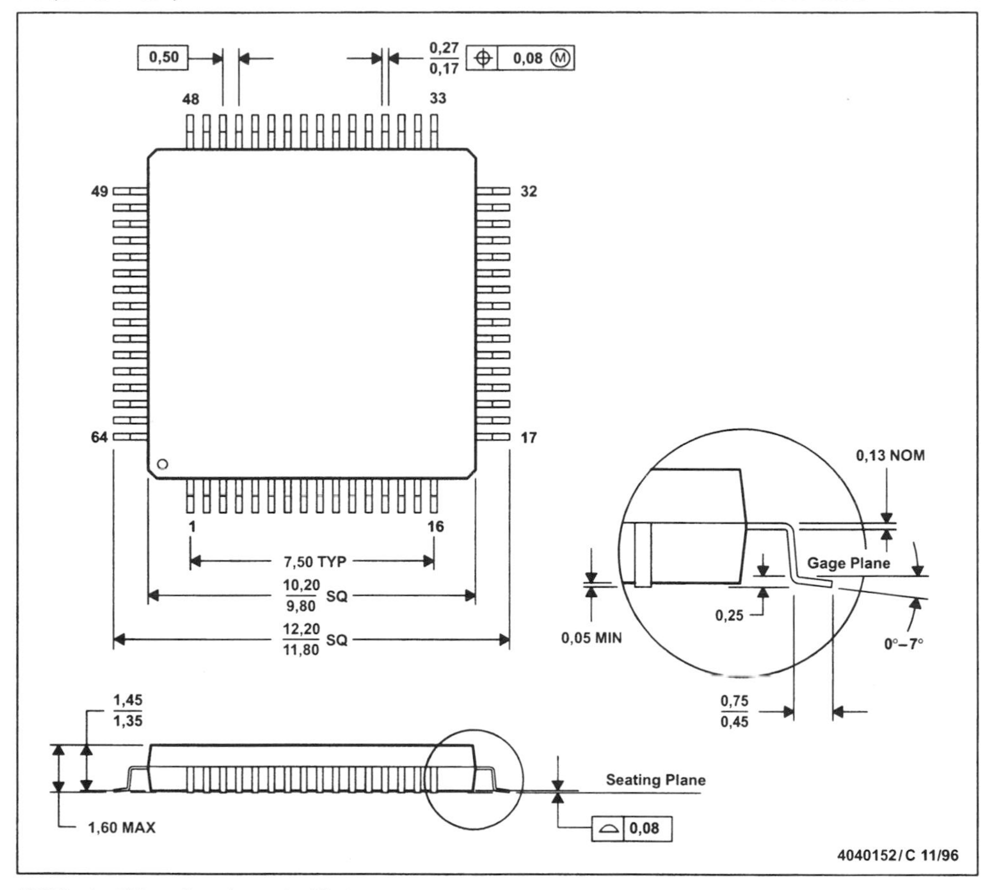

NOTES: A. All linear dimensions are in millimeters.
B. This drawing is subject to change without notice.
C. Falls within JEDEC MS-026
D. May also be thermally enhanced plastic with leads connected to the die pads.

MSP430x13x, MSP430x14x
MIXED SIGNAL MICROCONTROLLER

IMPORTANT NOTICE

Texas Instruments Incorporated and its subsidiaries (TI) reserve the right to make corrections, modifications, enhancements, improvements, and other changes to its products and services at any time and to discontinue any product or service without notice. Customers should obtain the latest relevant information before placing orders and should verify that such information is current and complete. All products are sold subject to TI's terms and conditions of sale supplied at the time of order acknowledgment.

TI warrants performance of its hardware products to the specifications applicable at the time of sale in accordance with TI's standard warranty. Testing and other quality control techniques are used to the extent TI deems necessary to support this warranty. Except where mandated by government requirements, testing of all parameters of each product is not necessarily performed.

TI assumes no liability for applications assistance or customer product design. Customers are responsible for their products and applications using TI components. To minimize the risks associated with customer products and applications, customers should provide adequate design and operating safeguards.

TI does not warrant or represent that any license, either express or implied, is granted under any TI patent right, copyright, mask work right, or other TI intellectual property right relating to any combination, machine, or process in which TI products or services are used. Information published by TI regarding third–party products or services does not constitute a license from TI to use such products or services or a warranty or endorsement thereof. Use of such information may require a license from a third party under the patents or other intellectual property of the third party, or a license from TI under the patents or other intellectual property of TI.

Reproduction of information in TI data books or data sheets is permissible only if reproduction is without alteration and is accompanied by all associated warranties, conditions, limitations, and notices. Reproduction of this information with alteration is an unfair and deceptive business practice. TI is not responsible or liable for such altered documentation.

Resale of TI products or services with statements different from or beyond the parameters stated by TI for that product or service voids all express and any implied warranties for the associated TI product or service and is an unfair and deceptive business practice. TI is not responsible or liable for any such statements.

Mailing Address:

Texas Instruments
Post Office Box 655303
Dallas, Texas 75265

Index

LIMITED WARRANTY AND DISCLAIMER OF LIABILITY

[[NEWNES.]] AND ANYONE ELSE WHO HAS BEEN INVOLVED IN THE CREATION OR PRODUCTION OF THE ACCOMPANYING CODE ("THE PRODUCT") CANNOT AND DO NOT WARRANT THE PERFORMANCE OR RESULTS THAT MAY BE OBTAINED BY USING THE PRODUCT. THE PRODUCT IS SOLD "AS IS" WITHOUT WARRANTY OF ANY KIND (EXCEPT AS HEREAFTER DESCRIBED), EITHER EXPRESSED OR IMPLIED, INCLUDING, BUT NOT LIMITED TO, ANY WARRANTY OF PERFORMANCE OR ANY IMPLIED WARRANTY OF MERCHANTABILITY OR FITNESS FOR ANY PARTICULAR PURPOSE. [[NEWNES.]] WARRANTS ONLY THAT THE MAGNETIC CD-ROM(S) ON WHICH THE CODE IS RECORDED IS FREE FROM DEFECTS IN MATERIAL AND FAULTY WORKMANSHIP UNDER THE NORMAL USE AND SERVICE FOR A PERIOD OF NINETY (90) DAYS FROM THE DATE THE PRODUCT IS DELIVERED. THE PURCHASER'S SOLE AND EXCLUSIVE REMEDY IN THE EVENT OF A DEFECT IS EXPRESSLY LIMITED TO EITHER REPLACEMENT OF THE CD-ROM(S) OR REFUND OF THE PURCHASE PRICE, AT [[NEWNES.]]'S SOLE DISCRETION.

IN NO EVENT, WHETHER AS A RESULT OF BREACH OF CONTRACT, WARRANTY OR TORT (INCLUDING NEGLIGENCE), WILL [[NEWNES.]] OR ANYONE WHO HAS BEEN INVOLVED IN THE CREATION OR PRODUCTION OF THE PRODUCT BE LIABLE TO PURCHASER FOR ANY DAMAGES, INCLUDING ANY LOST PROFITS, LOST SAVINGS OR OTHER INCIDENTAL OR CONSEQUENTIAL DAMAGES ARISING OUT OF THE USE OR INABILITY TO USE THE PRODUCT OR ANY MODIFICATIONS THEREOF, OR DUE TO THE CONTENTS OF THE CODE, EVEN IF [[NEWNES.]] HAS BEEN ADVISED OF THE POSSIBILITY OF SUCH DAMAGES, OR FOR ANY CLAIM BY ANY OTHER PARTY.

ANY REQUEST FOR REPLACEMENT OF A DEFECTIVE CD-ROM MUST BE POSTAGE PREPAID AND MUST BE ACCOMPANIED BY THE ORIGINAL DEFECTIVE CD-ROM, YOUR MAILING ADDRESS AND TELEPHONE NUMBER, AND PROOF OF DATE OF PURCHASE AND PURCHASE PRICE. SEND SUCH REQUESTS, STATING THE NATURE OF THE PROBLEM, TO ELSEVIER SCIENCE CUSTOMER SERVICE, 6277 SEA HARBOR DRIVE, ORLANDO, FL 32887, 1-800-321-5068. [[NEWNES.]] SHALL HAVE NO OBLIGATION TO REFUND THE PURCHASE PRICE OR TO REPLACE A CD-ROM BASED ON CLAIMS OF DEFECTS IN THE NATURE OR OPERATION OF THE PRODUCT.

SOME STATES DO NOT ALLOW LIMITATION ON HOW LONG AN IMPLIED WARRANTY LASTS, NOR EXCLUSIONS OR LIMITATIONS OF INCIDENTAL OR CONSEQUENTIAL DAMAGE, SO THE ABOVE LIMITATIONS AND EXCLUSIONS MAY NOT [[NEWNES.]] APPLY TO YOU. THIS WARRANTY GIVES YOU SPECIFIC LEGAL RIGHTS, AND YOU MAY ALSO HAVE OTHER RIGHTS THAT VARY FROM JURISDICTION TO JURISDICTION.

THE RE-EXPORT OF UNITED STATES ORIGIN SOFTWARE IS SUBJECT TO THE UNITED STATES LAWS UNDER THE EXPORT ADMINISTRATION ACT OF 1969 AS AMENDED. ANY FURTHER SALE OF THE PRODUCT SHALL BE IN COMPLIANCE WITH THE UNITED STATES DEPARTMENT OF COMMERCE ADMINISTRATION REGULATIONS. COMPLIANCE WITH SUCH REGULATIONS IS YOUR RESPONSIBILITY AND NOT THE RESPONSIBILITY OF **[[NEWNES.]]**.